Technology of Plant Biotechnology

TECHNOLOGY OF PLANT BIOTECHNOLOGY

Dr. Rayees Afzal Mir

NAMYA PRESS.com

Published by

Namya Press

India: 213, Vardaan House, 7/28 Ansari Road, Daryaganj, Delhi-110002

Email: namyapress@gmail.com

Website: https://namyapress.com

Edition	:	First Published in 2022 (Namya Press)
Title	:	Technology of Plant Biotechnology
Authors	:	Dr. Rayees Afzal Mir
ISBN	:	978-93-5545-131-6
Copyright	:	© 2022 All Rights Reserved

Contents

1

Technology of Plant Biotechnology

Dr. Rayees Afzal Mir

School of Agricultural Sciences, Glocal University, India

ABSTRACT

Plant biotechnology (PBT) incorporates a large number of logical instruments and procedures for screening and hereditary control of plants to foster helpful or valuable plant/plant items. The capability of these devices and methods could be increased by Nano mechanical intercessions. The oddity and ingenuity of the violating discipline of nanotechnology benchmarks the union, portrayal, control, and utilizations of issue at a size of 1-100 nm to such an extent that it deviates from its mass partners showing neoteric properties.

Nanotechnology has prepared crop biotechnologists to explain atomic pathways for plant reactions and designer architect crops by utilization of nanomaterials (NMs) as vehicles for quality or proteins. Enlarging NMs has prompted advancement of bionic plants that could proficiently catch sunlight based radiations for higher efficiency also as could be multiplexed to act as sensors. Nanoprobes like semiconductor and metal nanocrystals could be extremely valuable for high-goal imaging of the plant cell and organs.

These nanobioimaging devices could be tackled to observe elements of plant cell's biochemical and physiological reactions continuously, and NMs could be utilized for four-layered imaging of plant development reactions. The plant infection theranostics has additionally been hugely improved, with the goal that there is more noteworthy precision, speed, and responsiveness for location of hereditary qualities of nuisances, microbes, and yield plants including nanobarcodes for recognizable proof of hereditarily altered crops.

The plant cell nanofactories couldn't perform green and ecofriendly union of NMs yet additionally improve/modify seed germination and plant development through a plenty of complicated NM-organic framework connections. This part would see different nanotechnological mediations for PBT and would talk about the current status of innovative work in this best in class discipline.

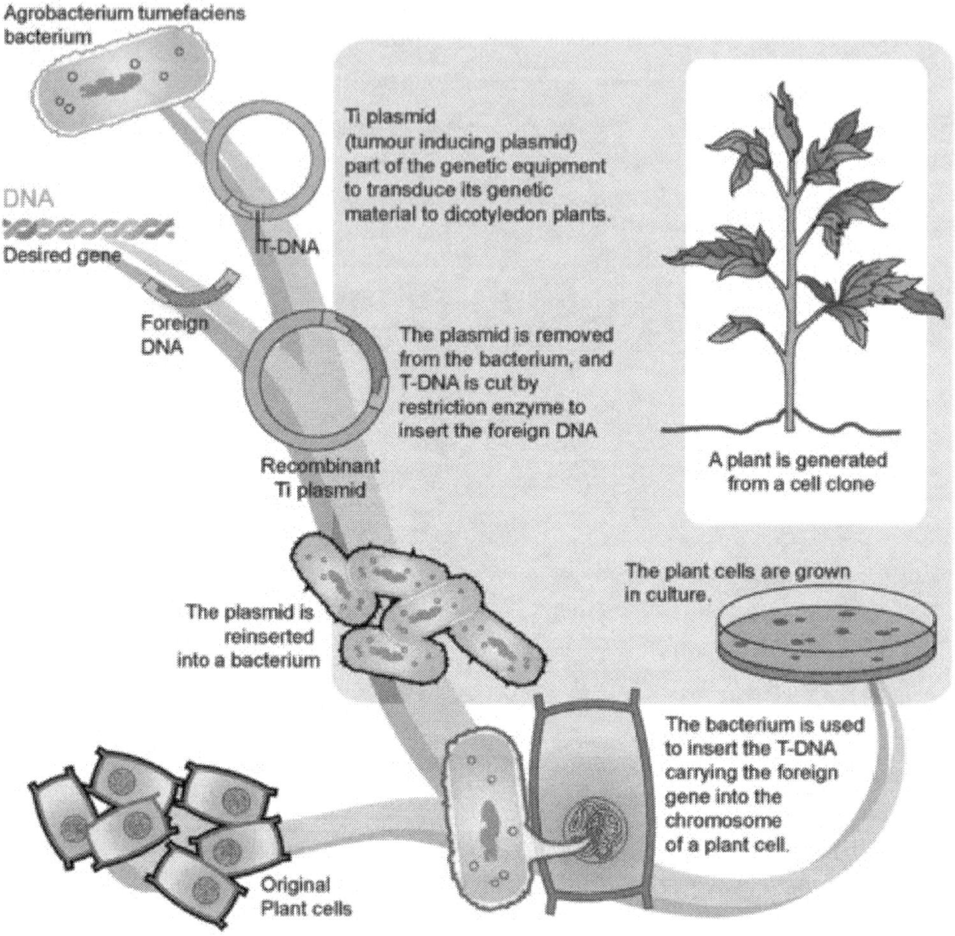

Introduction

Biotechnology comprehensively alludes to advancement of strategies for the use of natural standards to make items by organic specialists. For instance, the creation of wine by aging. The term biotechnology was begat by Karl Ereky, a Hungarian designer in 1919. The term plant biotechnology alludes to use of designing methods to change plants to help people.

In a limited definition, biotechnology alludes to hereditary control of creatures for explicit purposes. Biotechnology has two branches to be specific, tissue culture and hereditary designing. Tissue culture alludes to the in vitro development of cells, tissues, organs, incipient organisms, seeds and protoplasts on supplement media under aseptic circumstances in a research facility. The term hereditary designing alludes to control of creatures at the sub-atomic level straightforwardly including the DNA. Hereditary designing can alter bacterial cells to orchestrate totally new substances, for example chemicals, immunizations, monoclonal antibodies, and so on, or bring novel attributes into plants or creatures.

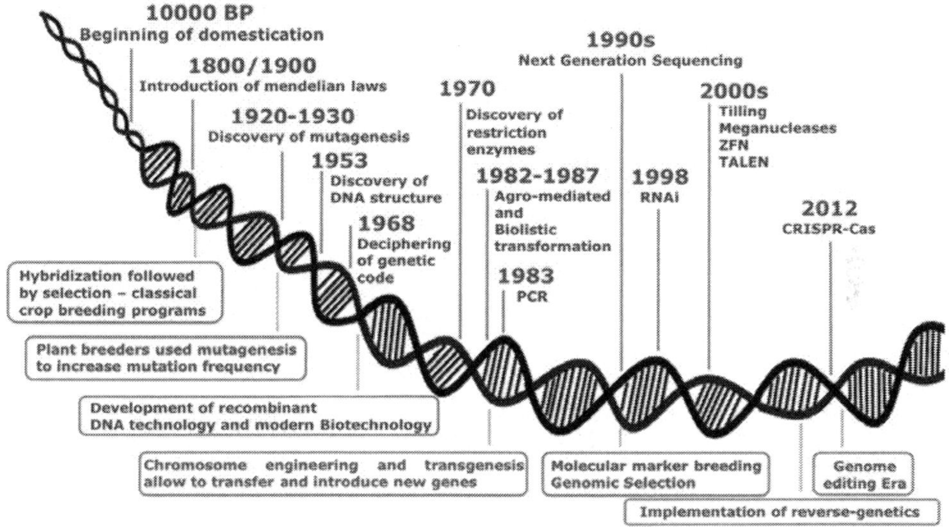

History of Plant Tissue Culture

Gottlieb Haberlandt is viewed as the Dad of plant tissue culture as he anticipated the totipotency of plant cells and endeavored in vitro culture of plant mesophyll cells as soon as 1902. Totipotency is the capacity of a plant cell to duplicate, separate and develop into a total plant. The principal undeveloped organism culture was endeavored by Hanning in 1904. In 1925, Laibach recuperated half breed descendants from an interspecific cross in Linum by involving zygotic undeveloped organisms of seeds as explant. The primary plant development chemical indoleacetic corrosive (IAA) was found in 1930s by F. Kogl et al. In 1934, Teacher Philip White effectively refined tomato roots on the medium, later called as White's medium. In 1939, Gautheret effectively refined carrot tissues and the chance of developing plant tissues for a limitless period was autonomously embraced by Gautheret, White and Nobecourt in 1939. In 1955 Mill operator and Skoog distributed their disclosure of the chemical kinetin, a cytokinin. The main plant from a developed plant cell was recovered by Braun in 1959. The focal point of the researchers later moved to planning of single cell societies. Muir (1953-54) exhibited that, callus tissues in fluid medium when exposed to shaking, broke into single cells. In 1960, Bergmann fostered the technique for cloning of these single cells by separating suspension societies. This procedure called Plating method is broadly utilized.

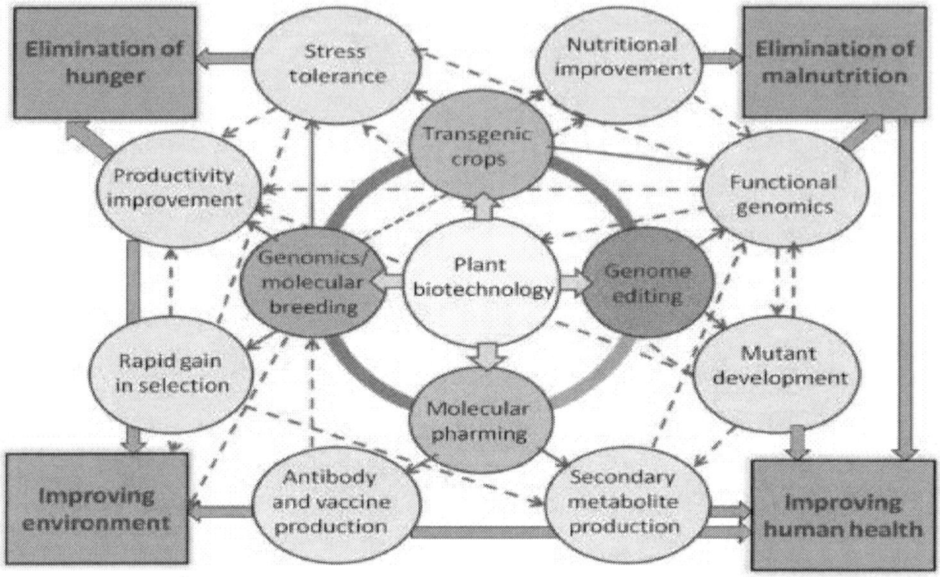

For instance, a solitary harvest might be expected to give reasonable food and energizing nourishment, security of the climate, and open doors for occupations and pay. Finding or creating reasonable plants is commonly a profoundly mind boggling challenge.

Plant biotechnologies that help with growing new assortments and characteristics incorporate hereditary qualities and genomics, marker-helped choice (MAS), and transgenic (hereditary designed) crops. These biotechnologies permit specialists to recognize and plan qualities, find their capabilities, select for explicit qualities in hereditary assets and reproducing, and move qualities for explicit attributes into plants where they are required. NIFA finances exploration, preparing, and augmentation for creating and involving biotechnologies for food and agribusiness. Areas of work incorporate, however not restricted to:

- Hereditary designs and instruments
- Strategies for transgenic biotechnology (otherwise called hereditary designing)
- Distinguishing proof of characteristics and qualities that can add to public and worldwide objectives for farming
- Plant genome arrangements; atomic markers, and bioinformatics

- Quality Altering/Genome Altering
- Manufactured Science
- Most open exploration on transgenic crops centers here and there around two general goals:
- Better comprehension of all parts of the transgenic/hereditary designing interaction, for improving effectiveness, accuracy, and legitimate articulation of the additional qualities or nucleic corrosive particles.
- A more extensive scope of helpful and important qualities, including complex characteristics.

Biotechnology Means

Any innovative application that utilizes natural frameworks, living organic entities, or subordinates thereof, to make or alter items or cycles for explicit use. Clean coal innovation implies any innovation, including advancements applied at the precombustion, ignition, or post burning stage, at a new or existing office which will accomplish critical decreases in air discharges of sulfur dioxide or oxides of nitrogen related with the use of coal in the age of power, or cycle steam which was not in that frame of mind as of November 15, 1990. Innovation implies all innovation, plans, formulae, calculations, methodology, strategies, revelations, processes, procedures, thoughts, skill, innovative work, specialized information, devices, materials, particulars, processes, developments (regardless of whether patentable or patentable and whether decreased to rehearse), mechanical assembly, manifestations, enhancements, works of creation in any media, classified, restrictive or nonpublic data, and other comparative materials, and all accounts, charts, drawings, reports, examinations and different compositions, and other unmistakable encapsulations of the previous in any structure whether recorded in this, for each situation, other than Programming.

Bioassay implies the assurance of sorts, amounts or fixations and, now and again, the areas of radioactive material in the human body, whether by direct estimation, in vivo counting, or by examination and assessment of materials discharged or eliminated from the human body. For motivations behind these standards, "radio bioassay" is an identical term. Clinical administration method implies a training which is utilized to control the expense or use of medical care administrations or physician endorsed drug use.

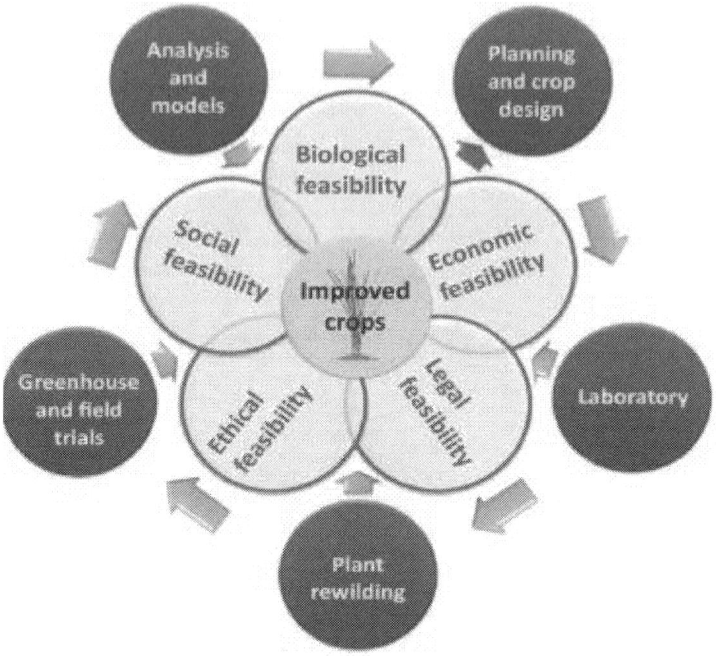

The term incorporates, without constraint, the utilization of step treatment, earlier approval or arranging medications and gadgets in light of cost, type or technique for organization. Cosmetology implies any one or potentially mix of practices by and large and generally performed by and known as the control of magnificence culturalism, cosmeticians, beauticians or stylists or any individual holding oneself out as rehearsing cosmetology in or upon a spot or premises. Cosmetology will incorporate, yet generally not be restricted to, the accompanying: adorning, orchestrating, dressing, twisting, waving, purifying, enhancing, trimming, burning, dying, shading, or comparative work upon the hair of any individual using any and all means and with hands or mechanical or electrical contraptions, gadgets or machines or by utilization of corrective arrangements, germicides, tonics, salves, creams etc., kneading, purging, invigorating, controlling, working out, embellishing or comparative work, the scalp, face, neck, arms, hands, bust or upper piece of the body, or manicuring, pedicuring or chiseling the nails of any individual. Clean coal innovation exhibition project implies a venture utilizing reserves appropriated under the heading "Branch of Energy—Clean Coal Innovation," up to an aggregate sum of $2,500,000,000 for business show of clean coal innovation, or comparable tasks financed through allotments for the Ecological Insurance Organization.

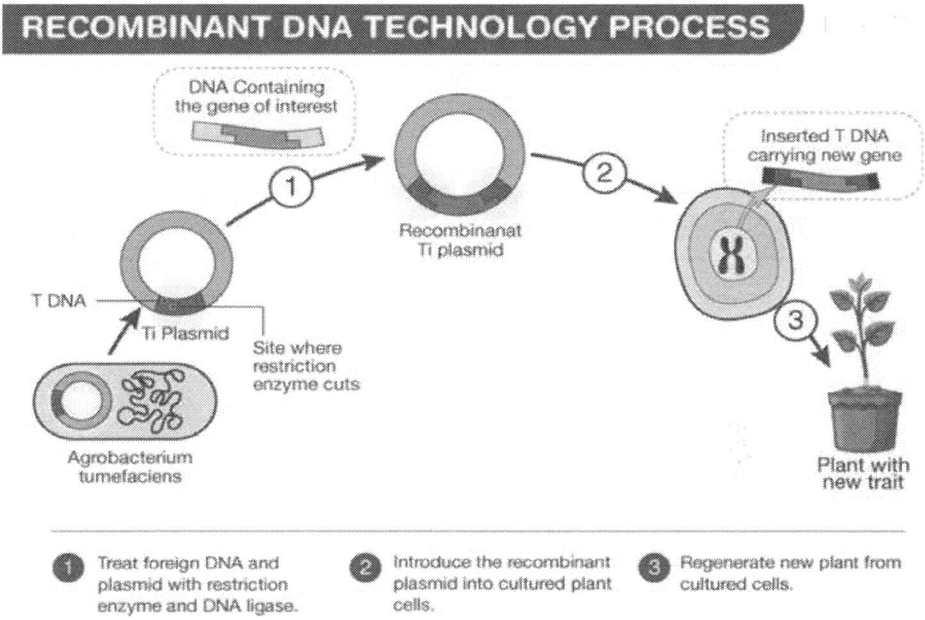

The government commitment for a passing venture will be something like 20% of the complete expense of the show project. Tetrahydrocannabinol implies the regular or engineered counterparts of the substances contained in the plant, or in the resinous extractives of, Weed sativa, or any manufactured substances, mixtures, salts, or subsidiaries of the plant or synthetics and their isomers with comparable compound design and pharmacological movement. Biomethane implies biogas that meets pipeline quality petroleum gas principles. New Innovation implies any creation, disclosure, improvement, or advancement, that was not accessible to the gaining organization on the viable date of the agreement, whether patentable, including, yet not restricted to, new cycles, arising innovation, machines, and enhancements to, or new utilizations of, existing cycles, machines, produces and programming. Additionally included are new PC projects, and enhancements to, or new uses of, existing PC projects, whether copyrightable and any new cycle, machine, including programming, and upgrades to, or new utilizations of, existing cycles, machines, produces and programming. High-innovation movement implies that term as characterized in area 3 of the Michigan monetary development authority act, 1995 Dad 24, MCL 207.803. nanomaterial implies a characteristic, coincidental or fabricated material containing particles in an unbound state or as a total or as an agglomerate and where, for 50 % or a greater amount of the particles in the number size conveyance, at least one outer aspects is in the size range 1-100 nm; Qualified high-innovation business implies a business that is both of the accompanying:

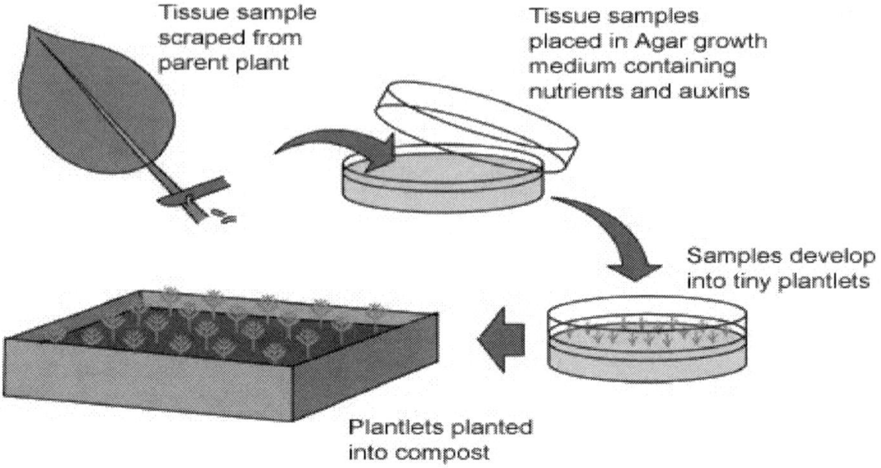

Assembling Innovation implies all possible licenses, patent applications, skill, and all licensed innovation privileges related therewith that are possessed or constrained by Licensor, and including all substantial epitomes thereof, that are important or valuable for the assembling of adeno-related infections, adeno-related infection vectors, exploration or business reagents related thereto, Authorized Items, or different items, including producing processes, specialized data connecting with the techniques for make, conventions, standard working methods, group records, examines, definitions, quality control information, determinations, increase, all possible enhancements, adjustments, and changes thereto, and all possible exercises related with such production. All science, assembling, and controls (CMC), drug ace documents (DMFs), or comparative materials gave to administrative specialists and the data contained in that are considered Assembling Innovation.

Radiopharmaceutical implies any medication that shows unconstrained crumbling of unsteady cores with the discharge of atomic particles or photons and incorporates any non-radioactive reagent unit or radionuclide generator that is expected to be utilized in the planning of any such substance, however does exclude medications, for example, carbon-containing mixtures or potassium-containing salts that incorporate follow amounts of normally happening radionuclides. The term likewise incorporates any natural item that is marked with a radionuclide or planned exclusively to be named with a radionuclide. Cropland implies land utilized for the creation of adjusted crops for reap, alone or in a turn with grasses and vegetables, and incorporates line crops, little grain crops, feed crops, nursery crops, plantation crops, and other comparable specialty crops.

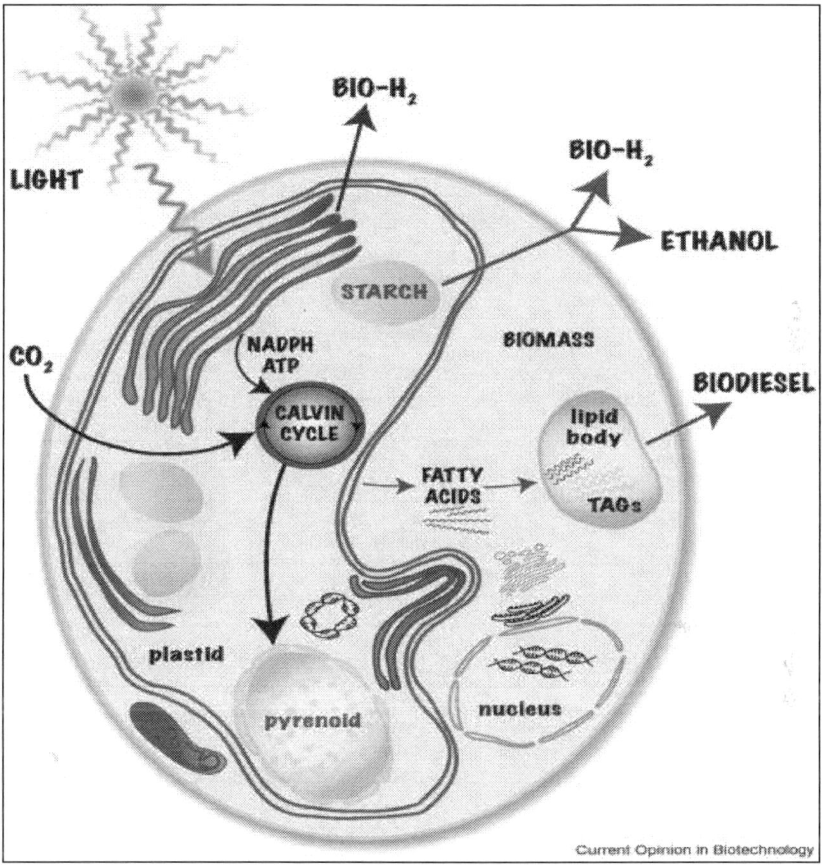

Weed implies all pieces of the plant Pot sativa Linnaeus, Marijuana indica, or Pot ruderals, regardless of whether developing; the seeds thereof; the sap, whether rough or cleansed, extricated from any piece of the plant; and each compound, produce, salt, subsidiary, combination, or arrangement of the plant, its seeds, or gum. "Pot" likewise implies the isolated pitch, whether rough or refined, acquired from marijuana. "Marijuana" does exclude the full grown stalks of the plant, fiber delivered from the stalks, oil or cake produced using the seeds of the plant, some other compound, make, salt, subsidiary, combination, or arrangement of the experienced stalks (with the exception of the tar separated consequently), fiber, oil, or cake, or the disinfected seed of the plant which is unequipped for germination.

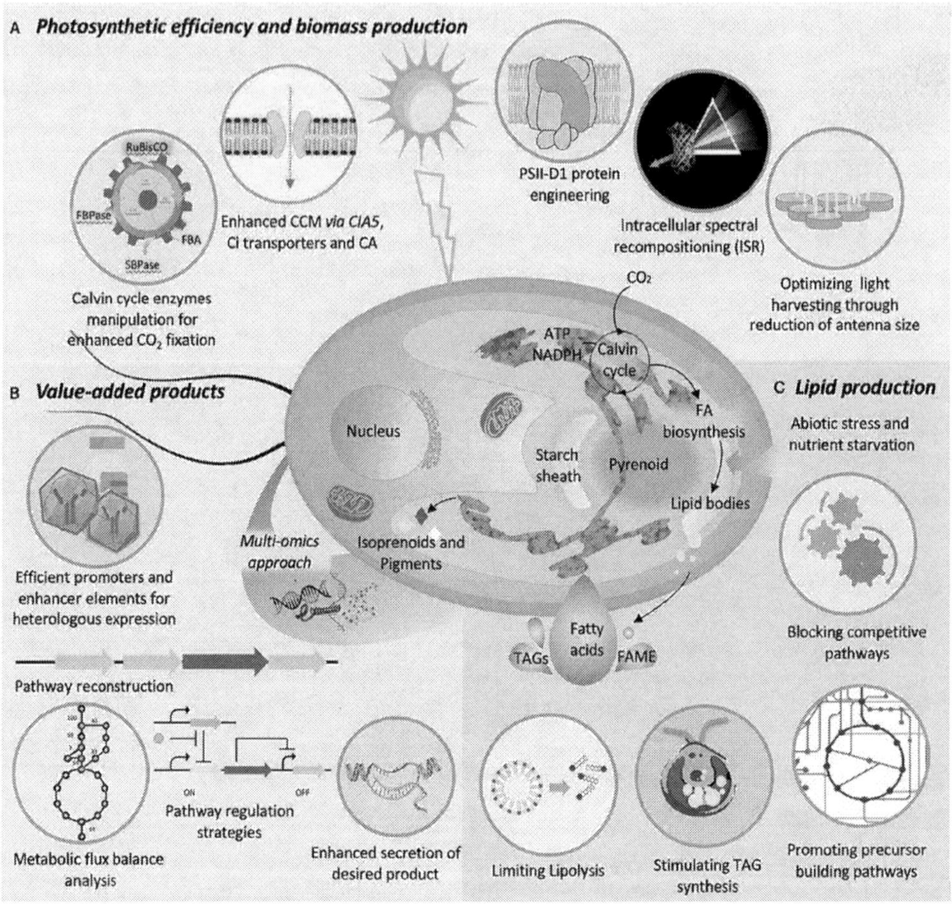

With the end goal of this division, "marijuana" doesn't imply "modern hemp" as characterized by Area 11018.5 of the Wellbeing and Security Code. Weed and the expression "cannabis" might be utilized conversely. destruction implies any of the accompanying demonstrations committed with expectation to obliterate, in entire or to some extent, a public, ethnical, racial or strict gathering, thusly: Gasification implies the sub stoichiometric oxidation or steam transformation of a substance to deliver a vaporous blend containing at least two of the accompanying: (I) oxides of carbon; (ii) methane; and (iii) hydrogen; Cooperation Innovation implies the Coordinated effort Skill and the Joint effort Licenses. Pot plant implies any plant of the class Pot; Assembling implies the creation, readiness, engendering, change or handling of a medication or gadget, either straightforwardly or in a roundabout way, by extraction from substances of normal beginning or freely through compound or natural blend, and incorporates any bundling or repackaging of the substances or naming or relabeling of its compartment, and the advancement and showcasing of such medications and gadgets. Fabricating additionally incorporates the readiness and advancement of monetarily accessible items from mass mixtures for resale by drug specialists, professionals, or different people.

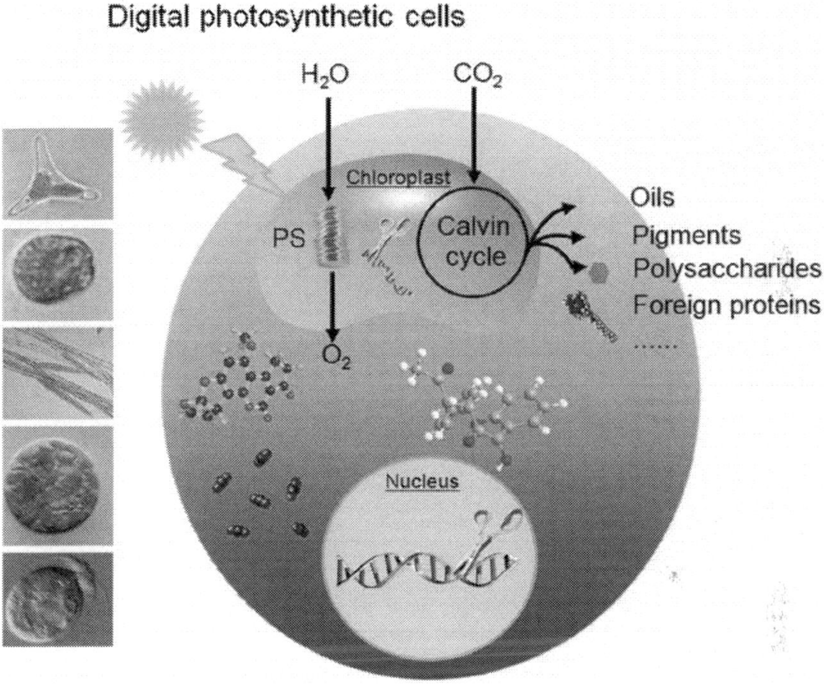

Cooperative medication treatment the board implies support by an approved drug specialist and a doctor in the administration of medication treatment as per a composed local area practice convention or a composed emergency clinic practice convention.

What is Biotechnology?

Biotechnology is the utilization of science to foster new items, strategies and living beings planned to work on human wellbeing and society. Biotechnology, frequently alluded to as biotech, has existed starting from the start of human progress with the training of plants, creatures and the revelation of maturation.

Early uses of biotech prompted the advancement of items like bread and antibodies. Nonetheless, the discipline has advanced fundamentally throughout the past hundred years in manners that control the hereditary designs and biomolecular cycles of living creatures. The advanced act of biotechnology draws from different disciplines of science and innovation, including the accompanying:

- atomic science
- science
- bionics
- hereditary designing
- genomics
- nanotechnology
- informatics

This approach has brought about developments and leap forwards in the accompanying regions:

- medications and therapeutics that forestall and treat sickness;
- clinical diagnostics, for example, pregnancy tests;
- biofuels that are manageable, decreasing waste and contamination; and
- hereditarily changed life forms (GMOs) that lead to more proficient and financially savvy farming.

Present day utilizations of biotechnology work most frequently through hereditary designing, which is otherwise called recombinant DNA innovation. Hereditary designing works by adjusting or connecting with the hereditary cell structures. Each cell in a creature or plant contains qualities that produce proteins. Those proteins decide the attributes of the organic entity.

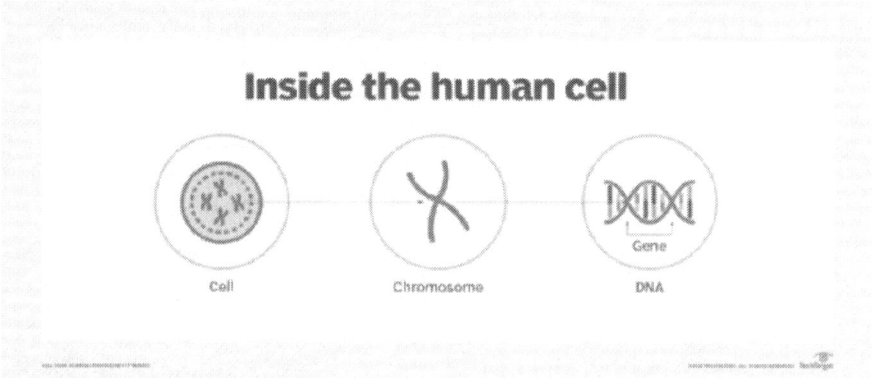

By changing or communicating with qualities, researchers can reinforce the attributes of a life form or make a totally new organic entity. These adjusted and new life forms might be helpful to people, like harvests with more significant returns or expanded protection from dry season. Hereditary designing likewise empowers the hereditary adjustment and cloning of creatures, two disputable turns of events.

Segments of DNA called genes are responsible for producing every characteristic in animals and plants.

History of Biotechnology

Biotechnology started somewhere around quite a while back with the farming upheaval. This early period was described by taking advantage of living creatures in their regular structures or adjusting their hereditary cosmetics through particular rearing.

Around similar time, people figured out how to bridle the natural course of aging to create bread, liquor and cheddar. Individuals likewise started changing the hereditary cosmetics of trained plants and creatures through specific rearing.

Particular rearing works by reproducing guardians with advantageous attributes to communicate or dispose of specific hereditary qualities in their posterity. Over the long haul, species that are specifically reared develop to be unique in relation to their wild progenitors. For example, during the farming upset, wheat was specifically reproduced to remain on its stem when collected as opposed to tumbling to the ground like wild wheat. Canines were specifically reared to be more tame than their wolf predecessors.

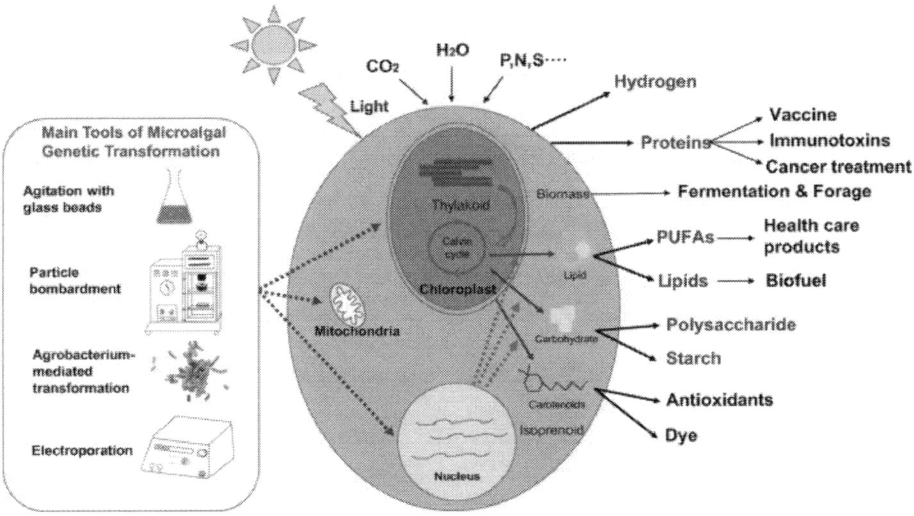

Notwithstanding, biotech techniques, for example, specific reproducing can consume a large chunk of the day to show changes in species. Biotechnology stayed restricted to these sluggish, agrarian techniques until the nineteenth century when researcher Gregor Mendel found the fundamental standards of heredity and hereditary qualities.

Additionally, during that period, researchers Louis Pasteur and Joseph Lister found the microbial cycles of aging. This established the groundwork for biotechnology enterprises where researchers cooperate all the more straightforwardly with the sub-atomic and hereditary cycles of creatures.

In view of crafted by these researchers, hereditary designing was created in 1973. This technique is the groundwork of current biotechnology practices and ongoing advances. It empowered the main direct control of plant and creature genomes, which is the finished arrangement of qualities present in a cell.

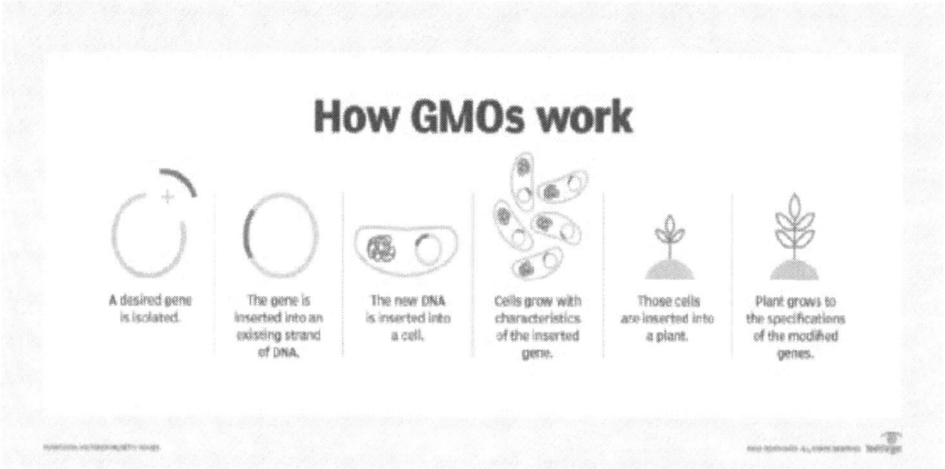

GMOs depend on genetic engineering for their creation and can be modified in any way a scientist chooses.

Over the last 100 hundred years or so, biotechnology emerged with the following discoveries and advancements:

1919. Hungarian scientist Karl Ereky coins the term *biotechnology*.

1928. Alexander Fleming discovers penicillin, the first true antibiotic.

1943. Oswald Avery proves DNA carries genetic information.

1953. James Watson and Francis Crick discover the double helix structure of DNA.

1960s. Insulin is synthesized to fight diabetes, and vaccines for measles, mumps and rubella are developed.

1969. The first synthesis of an enzyme in vitro, or outside the body, is conducted.

1973. Herbert Boyer and Stanley Cohen develop genetic engineering with the first insertion of DNA from one bacteria into another.

1980s. The first biotech drugs to treat cancer are developed.

1890. The United States Supreme Court rules that a "live human-made microorganism is patentable subject matter," meaning GMOs can be intellectual property.

1982. A biotech-developed form of insulin becomes the first genetically engineered product approved by the U.S. Food and Drug Administration (FDA).

1983. The first genetically modified plant is introduced.

1993. GMOs are introduced into agriculture with the FDA approval of growth hormones that produce more milk in cows.

1997. The first mammal is cloned.

1998. The first draft of the Human Genome Project is created, giving scientists access to over 30,000 human genes and facilitating research on treatment of diseases such as cancer and Alzheimer's.

2010. The first synthetic cell is created.

2013. The first bionic eye is created.

2020. MRNA vaccine and monoclonal antibody technology is used to treat the SARS-CoV-2 virus.

Types of Biotechnology

The science of biotechnology is broken down into subdisciplines that are color-coded based on common uses and applications.

- **Red biotechnology** involves medical processes, such as using organisms to produce new drugs and stem cells to regenerate damaged human tissues and grow and regrow entire organs.
- **White or gray** refers to industrial processes, such as the development of new chemicals or new biofuels for vehicles.
- **Green** covers agricultural processes, such as producing pest-resistant crops, disease-resistant animals and environmentally friendly agricultural practices.

- **Gold**, also known as bioinformatics, is a cross between biological processes and informatics. It refers to the methods healthcare workers use to gather, store and analyze biological data to treat patients.
- **Blue** encompasses processes in marine and aquatic environments, such as converting aquatic biomass into fuels and pharmaceuticals.
- **Yellow** refers to processes that aid food production, the most popular application being the fermentation of alcohol and cheese.
- **Violet** ensures the practice of biotechnology is in compliance with laws and ethical standards governing each field.
- **Dark** is the use of biotechnology for weapons or warfare.

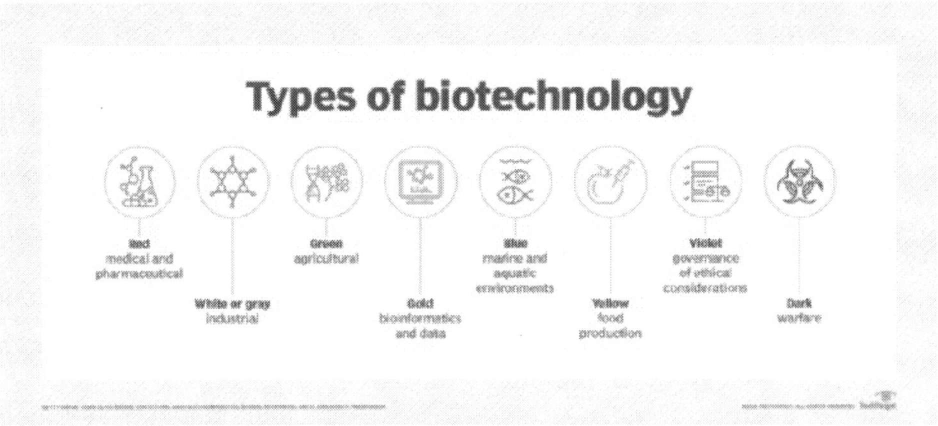

There are eight, color-coded categories of biotechnology.

Uses and Applications of Biotechnology

The use and commercialization of modern biotechnology often fall into four main fields: environment, medicine, industry and agriculture.

Environment

The aim of environmental biotechnology is to develop sustainable environmental practices that reduce pollution and waste. The following are examples of environmental biotech:

- **Phytoremediation** uses genetically engineered microorganisms to purify soils of heavy metals and other pollutants.
- **Bioremediation** introduces microorganisms into waste sites in order to organically break down no recyclable waste.
- **Plastic-eating bacteria** breaks down waste such as plastic in soils and water.

- **GMO foods** stay fresher longer and reduce food waste.
- **Genetic restoration** attempts to restore endangered species such as the American chestnut tree.
- **Cover-crops** such as corn are used as biofuels, replacing traditional fuel sources that produce greenhouse gas emissions when extracted and used.

Most genetic engineering is performed in laboratories.

Medicine

Medical biotechnology, also known as biopharma, aims to fight and prevent disease and improve healthcare. Biotechnology and biomedical research are the basis of the modern pharmaceutical industry. Uses include the following:

- stem cell research that helps replace or repair dead or defective cells;
- antibiotics development;
- gene therapies for diseases such as leukemia;
- research into dangerous pathogens and the antibodies that fight them;
- 3D printing or growing of organs and bones in labs; and
- mRNA vaccines, monoclonal antibody treatments and research for COVID-19.

Industry

Industrial biotechnology involves using microorganisms to produce industrial goods. Examples include the following:

- fermentation and the use of enzymes and microbes to streamline chemical manufacturing and reduce operational costs and chemical emissions;
- biofuels that use renewable crops such as corn to produce combustible fuel instead of natural, nonrenewable fossil fuel resources, such as petroleum and oil; and
- biodegradable garments and textiles made from the proteins of living organisms, such as the silk proteins of spiders.

Agriculture

Agricultural biotechnology genetically engineers plants and animals to produce more efficient agriculture, increase nutritional value and reduce food insecurity. Some examples of agricultural biotechnology are the following:

- biologically produced pesticides and herbicides that are less harmful to humans than chemical ones;
- drought-resistant crops;
- minimal space-resilient crops;
- meat grown in labs or using 3D printers;

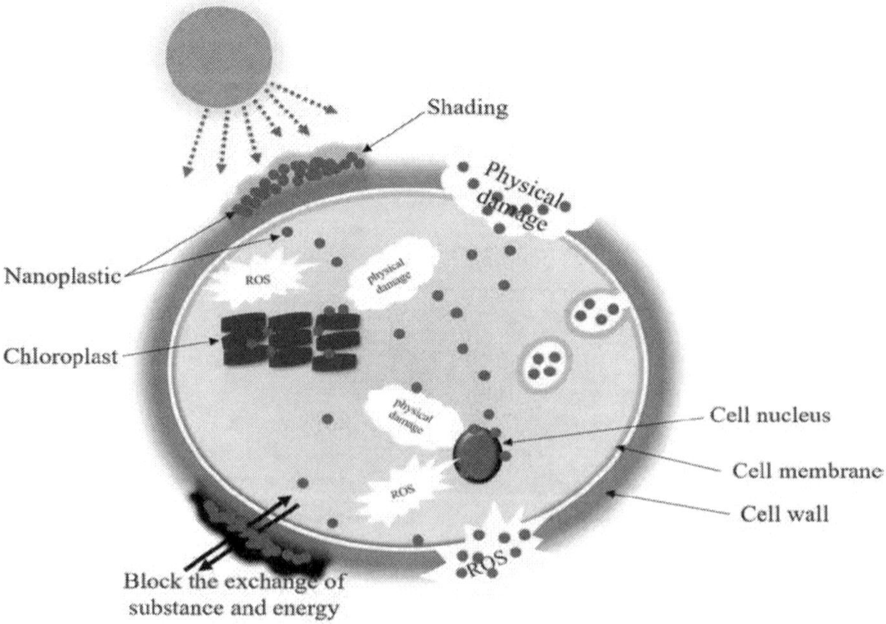

- gluten-free grains friendly to sufferers of celiac;
- selective breeding that produces healthier, bigger livestock and crops; and
- nutrient supplementation that infuses food with added nutrients to improve diets and medical treatments.

What are the Advantages of Biotechnology?

Biotechnology production offers a variety of advantages and solutions to critical problems. The main ones are the following:

- reducing pollution and waste to reverse catastrophic climate change and environmental damage;
- creating healthier, stronger and more-sustainable food products that boost nutrition and combat food insecurity;
- treating diseases in children before they are born by altering their genomes;
- designing medicine to boost the health and longevity of people, animals and plants; and
- cutting costs of farm supplies such as pesticides, while increasing crop yields and profits.

What are the Disadvantages of Biotechnology?

Biotechnology also comes with disadvantages and misuse. The main disadvantages include the following:

- **Biological warfare.** The potential exists for the development of pathogens and epidemics that could be used in a conflict zone to infect populations.
- **Decrease in biodiversity.** Mono-cropping or the farming of only a small number of genetically engineered crops could shrink the natural gene pool of species and make them less resilient and adaptable to sudden changes in environment.
- **Loss of soil fertility.** Bio-enhanced plants require more nutrients from soil and yield more crops. This can drain soil of fertile nutrients, devastate farmland and require the use of environmentally harmful fertilizers to make up for the nutrient shortfall.

- **High costs.** Biotechnology products often cost more than traditional products and have the potential to raise pricing structures in various industries.
- **Ethical considerations.** Gene manipulation raises a range of ethical issues, such as the genetic engineering of humans.
- **Safety questions.** Various groups have raised safety concerns about the health risks of GMOs and biotech-related medical developments, such as mRNA vaccines.

Concerns about biotechnology›s disadvantages have led to efforts to enact legislation restricting or banning certain processes or programs, such as human cloning, GMOs and embryonic stem-cell research.

Biotechnology is critical to environmentally sound advancements in agriculture. Learn more about how technology like artificial intelligence (AI) is improving the food industry.

This was last updated in November 2022

Continue Reading About Biotechnology (Biotech)

- Improving plant production with IT in the agriculture industry
- Open source machine learning accelerates winemaking
- Visual data analytics make genomics in healthcare possible
- The technology that enabled COVID-19 genome sequencing success
- IoT brings resource gains, sustainability to agriculture

History of Biotechnology

Individuals have been bridling natural cycles to work on their personal satisfaction for approximately 10,000 years, starting with the primary farming networks. Around quite a while back, people started to tap the organic cycles of microorganisms to make bread, cocktails, and cheddar and to protect dairy items. Yet, such cycles are not what is implied today by biotechnology, a term first generally applied to the sub-atomic and cell advances that started to arise during the 1960s and '70s. A juvenile "biotech" industry started to blend in the mid-to late 1970s, drove by Genentech, a drug organization laid out in 1976 by Robert A. Swanson and Herbert W. Boyer to market the recombinant DNA innovation spearheaded by Boyer, Paul Berg, and Stanley N. Cohen. Early organizations like Genentech, Amgen, Biogen, Cetus, and

Genex started by assembling hereditarily designed substances basically for clinical and natural purposes.

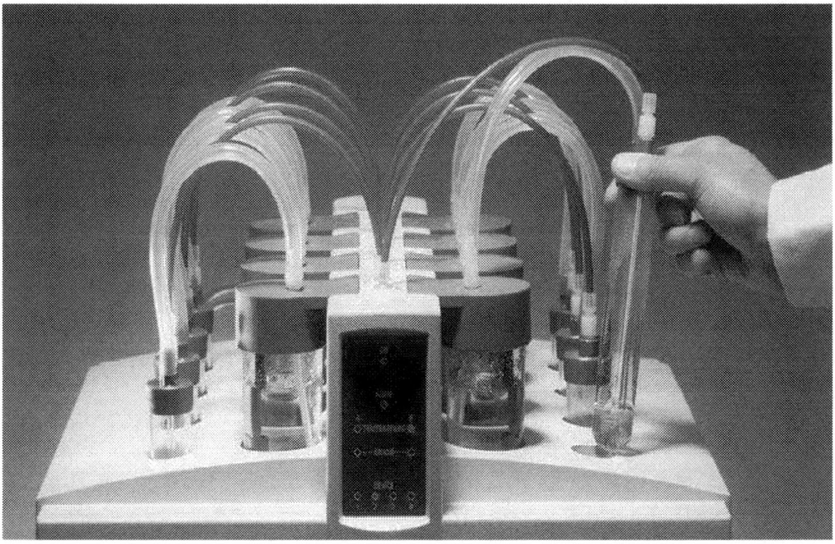

Biotechnology

For over 10 years, the biotechnology business was overwhelmed by recombinant DNA innovation, or hereditary designing. This procedure comprises of joining the quality for a valuable protein (frequently a human protein) into creation cells—like yeast, microbes, or mammalian cells in culture—which then start to deliver the protein in volume. During the time spent grafting a quality into a creation cell, another life form is made. From the get go, biotechnology financial backers and analysts were dubious about whether the courts would allow them to obtain licenses on organic entities; all things considered, licenses were not permitted on new organic entities that turned out to be found and distinguished in nature. In any case, in 1980, the U.S. High Court, on account of Jewel v. Chakrabarty, settled the matter by deciding that "a live human-made microorganism is patentable topic." This choice generated a rush of new biotechnology firms and the baby business' most memorable venture blast. In 1982 recombinant insulin turned into the primary item made through hereditary designing to get endorsement from the U.S. Food and Medication Organization (FDA). From that point forward, many hereditarily designed protein meds have been marketed all over the planet, including recombinant renditions of development chemical, thickening elements, proteins for animating the creation of red and white platelets, interferons, and cluster dissolving specialists.

Approaches and Instruments

In the early years, the principal accomplishment of biotechnology was the capacity to create normally happening helpful particles in bigger amounts than could be gotten from customary sources like plasma, creature organs, and human corpses. Recombinant proteins are likewise more averse to be tainted with microbes or to incite hypersensitive responses. Today, biotechnology specialists look to find the root sub-atomic reasons for infection and to mediate exactly at that level. In some cases this implies creating remedial proteins that expand the body's own provisions or that make up for hereditary lacks, as in the original of biotech prescriptions. (Quality treatment—inclusion of qualities encoding a required protein into a patient's body or cells—is a connected methodology.)

The biotechnology business has likewise extended its examination into the improvement of conventional drugs and monoclonal antibodies that postpone an illness. Fruitful creation of monoclonal antibodies was one of the main methods of biotechnology to arise during the last quarter of the twentieth hundred years. The particularity of monoclonal antibodies and their accessibility in amount have made it conceivable to devise delicate measures for a huge scope of naturally significant substances and to recognize cells from each other by distinguishing beforehand obscure marker particles on their surfaces. Such advances were made conceivable through the investigation of qualities (genomics), the proteins that they encode (proteomics), and the bigger organic pathways where they act.

Applications of Biotechnology

Biotechnology has various applications, especially in medication and horticulture. Models remember the utilization of biotechnology for consolidating natural data with PC innovation (bioinformatics), investigating the utilization of minuscule hardware that can enter the human body (nanotechnology), and potentially applying strategies of foundational microorganism examination and cloning to supplant dead or blemished cells and tissues (regenerative medication). Organizations and scholarly research facilities coordinate these unique advancements with an end goal to investigate descending into particles and furthermore to orchestrate vertically from atomic science toward substance pathways, tissues, and organs.

As well as being utilized in medical services, biotechnology has demonstrated accommodating in refining modern cycles through the disclosure and creation of natural compounds that flash synthetic responses (impetuses); for ecological cleanup, with catalysts that digest pollutants into innocuous synthetic compounds and afterward bite the dust in the wake of eating the accessible "food supply"; and in agrarian creation through hereditary designing.

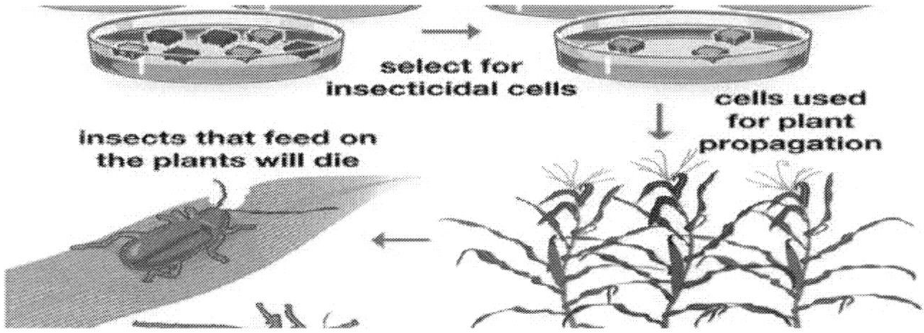

Genetically Modified Organisms

Agrarian uses of biotechnology have demonstrated the most disputable. A few activists and shopper bunches have called for prohibitions on hereditarily changed organic entities (GMOs) or for naming regulations to supply educate customers regarding the developing presence of GMOs in the food. In the US, the presentation of GMOs into farming started in 1993, when the FDA endorsed cow-like somatotropin (BST), a development chemical that lifts milk creation in dairy cows. The following year, the FDA endorsed the principal hereditarily changed entire food, a tomato designed for a more drawn out timeframe of realistic usability. From that point forward, administrative endorsement in the US, Europe, and somewhere else has been won by many farming GMOs, including crops that produce their own pesticides and harvests that endure the use of explicit herbicides used to kill weeds.

Concentrates on by the Assembled Countries, the U.S. Public Institute of Sciences, the European Association, the American Clinical Affiliation, U.S. administrative offices, and different associations have viewed GMO food sources as protected, yet cynics battle that making a decision about the drawn out wellbeing and environmental impacts of such crops is still too soon. In the late twentieth and mid 21st hundreds of years, the land region established in hereditarily altered crops expanded decisively, from 1.7 million hectares (4.2 million sections of land) in 1996 to 180 million hectares (445 million sections of land) by 2014. By 2014-15 around 90 percent of the corn, cotton, and soybeans established in the US were hereditarily changed. Most of hereditarily adjusted crops were filled in the Americas.

Generally, the incomes of U.S. what's more, European biotechnology businesses generally multiplied over the five-year time span from 1996 through 2000. Quick development went on into the 21st 100 years, filled by the presentation of new items, especially in medical services. By 2020 the biotechnology market size was assessed at $752.88 billion around the world, with new open doors for development arising specifically from government-and industry-driven endeavors to speed up drug improvement and item endorsement processes.

Plant Biotechnology: Areas and Scope

Advertisements:

In this article we will examine about the areas and extent of plant biotechnology.

The term 'Biotechnology' was begat in 1917 by a Hungarian Designer, Karl Ereky, to depict a cycle for enormous scope creation of pigs. The word biotechnology demonstrates the cooperation between the study of science and innovation. Biotechnology has been extensively characterized as, 'the turn of events and usage of organic cycles, structures and frameworks for getting greatest advantages to human and different types of life'.

As per the definition embraced by the European League of Biotechnology, made in 1978, 'Biotechnology makes it conceivable, through a coordinated utilization of information and strategies of natural chemistry, microbial science, hereditary qualities and synthetic engineering, to draw benefit, at the mechanical level, from the properties and limits of microorganisms and cell societies.

Notices:

Biotechnology offers the chance of creating, from generally accessible inexhaustible assets, substances and mixtures crucial for life.

Current biotechnology empowers a creature to deliver an absolutely new item which the living being doesn't or can't deliver regularly through the fuse of the technology of 'Hereditary designing'. The information on toddler intensity of plant cell, various parts of plant tissue culture strategies, cell combination methods alongside the utilization of rDNA innovation has fostered another area of 'Plant biotechnology'.

Regions and Parts of Plant Biotechnology:

Numerous areas of biotechnology have created through the association between various parts of science and designing; organic chemistry, biophysics, cell science, science, embryology, environment, microbial science, hereditary qualities, atomic science and immunology—this large number of disciplines of science and on the designing side substance and biochemical Engi-neering have assisted with fostering the various regions and parts of current biotechnology.

Table : Areas and Branches of Biotechnology along with their products

Areas/Brunches		Products
1.	Recombinant DNA Techniques (Genetic Engineering)	Fine chemicals, Enzymes, Vaccines, Hormones. Anti-bodies. Antibiotics
2.	Plant Cell Culture/Animal Cell Culture Techniques	Fine chemicals (alkaloids. sterokls, essential oils. dyes). Somatic embryos, Monoclonal antibodies. Diagnostic and Therapeutic kits
3.	Process Engineering	Harvesting. Pretreatment. Filtration. Effluent treatment. Water recycling. Product extraction Novel reactors
4.	Was Product Treatment and Illilization	Herbicideffn sect detoxification, Sewage treatment and utilization. Byproduct utilization
5.	Traditional Fermentation	Alcohol. Organic acids. Amino ac ids. Polysaccharides, Antibiotics
6.	Biofuels	Methane (bjogas front wastes). Hydrogen, Alcohol (from waste)
7.	Enzymes and Biaxatalysts	Food processing. Fine chemicals. Diagnostic kits.
8.	Nitrogen Fixation	Chemotherapy. B i osensors Microbial or Algal biofertilizer.

Recombinant DNA Techniques and Genetic Engineering:

Advertisements:

The knowledge of molecular biology has provided the ability to extract a gene coding for a desired product and transfer it to another organism which has helped in production of useful proteins, hormones, vaccines, enzymes, etc. by some microorganisms, plant cells or animal cell cultures.

The use of this technique is essential for four reasons:

First, large scale production of desired product;

Second, supply of natural source may be limited;

Third, the purity of naturally available material is doubtful;

Fourth, is the cost of the natural product may be higher.

Plant Cell Culture Techniques:

The capacity to culture plant cells for a huge scope, either for the creation of biomass or to remove the ideal item from cell societies (high yielding), has turned into an exceptionally beneficial innovation. Plant cells are important and rich wellspring of high worth medications and numerous optional metabolites. With the progression of this innovation the creation turns out to be more financial to deliver minimal expense and high volume materials.

Notices

Process Designing:

Research facility researchers can clone a clever quality, find another anti-toxin or catalyst catalysed process however it expects to move the information to the size of activity expected to make a valuable item in critical amount.

Reaping, pre-treatment, filtration of the natural substance, reactor plan and their control, recuperation/reuse of biocatalyst or creature, item extraction and investigation—every one of these need the support of synthetic or biochemical designers.

Squander Treatment and Usage:

Commercials:

The treatment of sewage and waste water from homegrown and modern sources is the biggest biotechnological industry. Squanders and by-products of exercises in agribusiness, ranger service and food industry can be utilized for different purposes, especially to create energy. Many kinds of squanders are reasonable for biological treatment and some might try and yield valuable items.

But anaerobic or oxygen consuming waste treatment, biodegradation is the interaction by which squander materials, for example, oil slick, herbicides, pesticides, and so on are debased by the activity of microbial frameworks.

Conventional Maturation:

Customarily, maturation has implied the development of consumable liquor from sugars. In any case, maturation—the use of microbial digestion to change basic unrefined components into significant items—can create valuable substances for example citrus extract, anti-toxins, biopolymer and single cell proteins. The capability of this cycle is interminable and changed as microorganisms themselves.

Advertisements:

Biofuels:

Biotechnology can give two arrangements towards the issues of fills in our present reality—one is new sort of energizes, another is elective carbon feedstock. Sugarcane waste can be utilized as hotspot for alcoholic fills. Creature ooze can be utilized for bio-gas creation, some of amphibian spices (Euchronia) are likewise utilized for biogas creation—which is elective wellspring of fills.

The most stylishly satisfying biotechnological fuel would be hydrogen gotten from the bio-photolysis of water, which depends on the combination of the photograph frameworks of plants with bacterial-inferred hydrogenase proteins and light.

Chemicals and Bio-catalysis:

Ads:

As of late extraordinary interest has been stirred in broadening the utilization of catalysts in food handling, compound creation, scientific and demonstrative frameworks, and in the treatment of sickness.

This cutting edge interest is generally a consequence of the more noteworthy comprehension of the capability and nature of compounds and furthermore the new technologies engaged with the treatment of biocatalysts, especially their immobilization so they might be fixed in a reactor or on a sensor test.

Bio-compost:

Promotions:

Living plants utilized as bio-composts have turned into a method for supplementing the accessibility of the scant wellspring of synthetic manures and furthermore improving the dirt. Leguminous plants improve the fruitfulness of soil because of presence of Rhizobium, other bio-composts are Azotobacter, Azospirillum and Azolla; other phosphate solubilizing microorganisms and mycorrhizal organisms have likewise tracked down a spot in bio-manure technology.

Degree and Possibilities of Plant Biotechnology:

Every single life form carries out its role inside its ideal cutoff points however the energy of present day biotechnology lies in the way that the use of various logical strategies (like hereditary designing) can improve the innate capacities of production considerably more too as supernatural occurrence can happen i.e., microorganisms can deliver mammalian chemical, and furthermore plant cell can create plastic granules.

Biotechnology in Horticulture:

Biotechnology is making another ground in the food or horticulture region. Different plant cell, tissue and organ culture procedures can help in miniature spread of various types of plants for various purposes, phytomass or biomass creation to get a few alkaloids or optional metabolites, anther societies are utilized for haploid creation, gametic and physical cell/ tissue societies are utilized for gametoclonal or somaclonal variety or for creation of fake seeds.

Hereditary designing and change procedures have helped underway of transgenic plants conveying advantageous characteristics like sickness opposition, bug obstruction, herbicide obstruction and furthermore might be utilized for expanding photosynthetic productivity, nitrogen fixing capacity, further developed stockpiling proteins, half breed crops, and so on.

Protoplast culture and combination have helped underway of physical crossovers between physically contrary species, allowing move of positive attributes from wild or unrelated crop species to our developed yield plants.

Biotechnology in Industry and Medication:

Promotions:

Modern microbial science is one more region getting significant consideration of biotechnologists; numerous drug medications and synthetics are presently being created in microbial cells in bigger amount using procedures of genetic designing. In the field of medication, insulin and interferon combined by microbes are as of now on the lookout.

Countless antibodies for inoculation against lethal diseases, DNA tests and monoclonal antibodies for analysis of different infections, human development chemical and other drug drugs for treatment of sicknesses are being produced through organisms. Quality treatment for a few deadly illnesses and DNA fingerprinting or autoantibody fingerprinting methods are extraordinary help in clinical and scientific science.

Biotechnology and Climate:

Biotechnology is additionally being utilized for managing ecological issues like contamination control, consumption of normal assets for non-renewable energy, reclamation of debased lands, biodiversity preservation. Bio-pesticides, bio-composts and biosensors (for location of metals), cleaning of metals and spilled oils, bio-checking in various businesses are the consequence of biotechnological improvements.

Biotechnology in Protein and Metabolic Designing:

Protein designing, metabolic designing are likewise the as of late evolved significant areas of biotechnology. Protein designing assists the creation of unrivaled catalysts and capacity proteins with the assistance of PC supported protein demonstrating.

Biotechnology research has helped in the production of metabolites at the modern scale by the utilization of PC supported presentation of various qualities and administrative components called metabolic designing. In the space of bioinformatics as of late data sets are accessible through data network all through the world which help the advancement of bio-mechanical examination in numerous areas.

2

Observation of Biotechnology

Dr. Rayees Afzal Mir
School of Agricultural Sciences, Glocal University, India.

Introduction

The expression "Biotechnology" was first authored in 1919 by Karl Ereky which implies items are created from unrefined substances with the guide of living life forms. Biotechnology isn't new. Man has been controlling living things to tackle issues and work on his lifestyle for centuries. Early horticulture focused on delivering food. Plants and creatures were specifically reared and microorganisms were utilized to make food things like drinks, cheddar and bread. The late eighteenth hundred years and the start of the nineteenth century saw the approach of immunizations, crop turn including leguminous yields and creature drawn apparatus. The finish of the nineteenth century was an achievement of science. Microorganisms were found, Mendel's work on hereditary qualities was achieved and initiates for examining maturation and other microbial cycles were laid out by Koch, Pasteur and Lister.

Biotechnology toward the start of the 20th century started to unite industry and agribusiness. During The Second Great War, maturation processes were fostered that delivered CH3)2CO from starch and paint solvents for the quickly developing car industry. Work during the 1930s was equipped towards utilizing surplus horticultural items to supply industry rather than imports or petrochemicals. The coming of The Second Great War brought the production of penicillin. The biotechnical center moved to drugs. The "cool conflict" years were overwhelmed by work with microorganisms in anticipation of natural fighting as well as anti-infection agents and maturation processes (Goodman, 1987).

Biotechnology is as of now being utilized in numerous areas including agribusiness, bioremediation, and food handling and energy creation. DNA fingerprinting is turning into a typical practice in criminology. Creation of insulin and different drugs is achieved through cloning of vectors that presently convey the picked quality. Immunoassays are utilized not just in medication for drug level and pregnancy testing, yet additionally by ranchers to help with identification of risky degrees of pesticides, herbicides and poisons on crops and in animal items. These examines likewise give fast field tests to modern synthetics in ground water, residue and soil. In horticulture, hereditary designing is being utilized to deliver plants that are impervious to bugs, weeds and plant sicknesses.

History of Biotechnology

Individuals have been tackling organic cycles to work on their personal satisfaction for nearly 10,000 years, starting with the principal agrarian networks. Around quite a while back, people started to tap the natural cycles of microorganisms to make bread, cocktails, and cheddar and to protect dairy items. Yet, such cycles are not what is implied today by biotechnology, a term first generally applied to the sub-atomic and cell innovations that started to arise during the 1960s and '70s. A youngster "biotech" industry started to combine in the mid-to late 1970s, drove by Genentech, a drug organization laid out in 1976 by Robert A. Swanson and Herbert W. Boyer to popularize the recombinant DNA innovation spearheaded by Boyer, Paul Berg, and Stanley N. Cohen. Early organizations like Genentech, Amgen, Biogen, Cetus, and Genex started by assembling hereditarily designed substances fundamentally for clinical and ecological purposes.

Biotechnology

For over 10 years, the biotechnology business was overwhelmed by recombinant DNA innovation, or hereditary designing. This method comprises of grafting the quality for a helpful protein (frequently a human protein) into creation cells—like yeast, microorganisms, or mammalian cells in culture—which then start to deliver the protein in volume. During the time spent grafting a quality into a creation cell, another living being is made. From the start, biotechnology financial backers and analysts were unsure about whether the courts would allow them to gain licenses on creatures; all things considered, licenses were not permitted on new life forms that turned out to be found and recognized in nature. Be that as it may, in 1980, the U.S. High Court, on account of Precious stone v.

Chakrabarty, settled the matter by deciding that "a live human-made microorganism is patentable topic." This choice generated an influx of new biotechnology firms and the baby business' most memorable speculation blast. In 1982 recombinant insulin turned into the main item made through hereditary designing to get endorsement from the U.S. Food and Medication Organization (FDA). From that point forward, many hereditarily designed protein drugs have been marketed all over the planet, including recombinant adaptations of development chemical, coagulating factors, proteins for animating the creation of red and white platelets, interferons, and cluster dissolving specialists.

Approaches and Tools

In the early years, the fundamental accomplishment of biotechnology was the capacity to deliver normally happening helpful atoms in bigger amounts than could be gotten from customary sources like plasma, creature organs, and human dead bodies. Recombinant proteins are additionally more averse to be defiled with microbes or to incite unfavorably susceptible responses. Today, biotechnology specialists try to find the root sub-atomic reasons for illness and to intercede unequivocally at that level. Once in a while this implies creating restorative proteins that expand the body's own provisions or that make up for hereditary lacks, as in the original of biotech meds. (Quality treatment—inclusion of qualities encoding a required protein into a patient's body or cells—is a connected methodology.)

The biotechnology business has likewise extended its investigation into the improvement of customary drugs and monoclonal antibodies that postpone an illness. Effective creation of monoclonal antibodies was one of the main procedures of biotechnology to arise during the last quarter of the twentieth hundred years. The explicitness of monoclonal antibodies and their accessibility in amount have made it conceivable to devise delicate measures for a colossal scope of organically significant substances and to recognize cells from each other by distinguishing beforehand obscure marker particles on their surfaces. Such advances were made conceivable through the investigation of qualities (genomics), the proteins that they encode (proteomics), and the bigger organic pathways where they act.

Utilizations of Biotechnology

Biotechnology has various applications, especially in medication and horticulture. Models remember the utilization of biotechnology for consolidating natural data with PC innovation (bioinformatics), investigating the utilization of tiny hardware that can enter the human body (nanotechnology), and potentially applying methods of immature microorganism examination and cloning to supplant dead or inadequate cells and tissues (regenerative medication). Organizations and scholastic research facilities coordinate these unique advances with an end goal to examine descending into particles and furthermore to incorporate vertical from atomic science toward substance pathways, tissues, and organs.

As well as being utilized in medical services, biotechnology has demonstrated supportive in refining modern cycles through the disclosure and creation of organic proteins that flash substance responses (impetuses); for ecological cleanup, with compounds that digest pollutants into innocuous synthetics and afterward kick the bucket in the wake of devouring the accessible "food supply"; and in rural creation through hereditary designing.

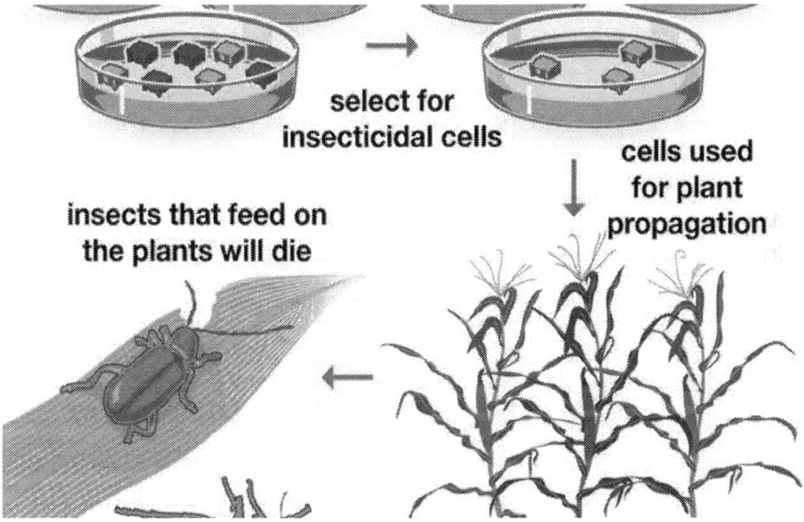

Genetically Modified Organisms

Rural utilizations of biotechnology have demonstrated the most dubious. A few activists and purchaser bunches have called for prohibitions on hereditarily changed living beings (GMOs) or for naming regulations to supply educate customers regarding the developing presence of GMOs in the food. In the US, the presentation of GMOs into farming started in 1993, when the FDA supported cow-like somatotropin (BST), a development chemical that lifts milk creation in dairy cows. The following year, the FDA endorsed the principal hereditarily changed entire food, a tomato designed for a more drawn out time span of usability. From that point forward, administrative endorsement in the US, Europe, and somewhere else has been won by many horticultural GMOs, including crops that produce their own pesticides and harvests that endure the use of explicit herbicides used to kill weeds.

Concentrates on by the Unified Countries, the U.S. Public Institute of Sciences, the European Association, the American Clinical Affiliation, U.S. administrative offices, and different associations have viewed GMO food varieties as protected, yet doubters battle that passing judgment on the drawn out wellbeing and natural impacts of such crops is still too soon. In the late twentieth and mid 21st hundreds of years, the land region established in hereditarily changed crops expanded decisively, from 1.7 million hectares (4.2 million sections of land) in 1996 to 180 million hectares (445 million sections of land) by 2014. By 2014-15 around 90 percent of the corn, cotton, and soybeans established in the US were hereditarily altered. Most of hereditarily altered crops were filled in the Americas.

In general, the incomes of U.S. also, European biotechnology enterprises generally multiplied over the five-year time frame from 1996 through 2000. Quick development went on into the 21st hundred years, filled by the presentation of new items, especially in medical care. By 2020 the biotechnology market size was assessed at $752.88 billion around the world, with new open doors for development arising specifically from government-and industry-driven endeavors to speed up drug advancement and item endorsement processes.

Biotechnology is multidisciplinary field which significantly affects our lives. The innovation is known since years which includes working with cells or cell-inferred particles for different applications. It has extensive variety of purposes and is named "innovation of trust" which influence human wellbeing, prosperity of other living things and our current circumstance. It has upset diagnostics and therapeutics; in any case, the significant difficulties to the people have been dangers acted by lethal infection diseases like avian influenza, Chikungunya, Ebola, Flu A, SARS, West Nile, and the most recent Zika infection. Customized medication is progressively perceived in medical care framework. In this section, the perusers would comprehend the utilizations of biotechnology in human medical care framework. It has likewise influenced the climate which is stacked by poisonous mixtures because of human industrialization and urbanization. Bioremediation process uses utilization of normal or recombinant creatures for the cleanup of ecological poisonous contaminations.

The advancement of bug and nuisance safe harvests and herbicide open minded crops has extraordinarily diminished the natural heap of harmful insect poisons and pesticides. The expansion in crop efficiency for taking care of world food and feed issue is tended to in agrarian biotechnology. The mechanical headways have zeroed in on advancement of substitute, inexhaustible, and feasible energy hotspots for creation of biofuels. Marine biotechnology investigates the items which can be gotten from sea-going life forms. Similarly as with each exploration region, the area of biotechnology is related with numerous moral issues and concealed fears. These are significant in characterizing regulations administering the possibility and endorsement for the lead of specific exploration.

The expression "biotechnology" was begat by a Hungarian specialist Karl Ereky, in 1919, to allude to the science and strategies that license items to be created from unrefined components with the guide of living organic entities. Biotechnology is a different field which includes either working with living cells or utilizing particles got from them for applications situated toward human government assistance utilizing shifted kinds of instruments and innovations. It is a mixture of natural science with designing by which living creatures or cells or parts are utilized for creation of items and administrations. The primary subfields of biotechnology are clinical (red) biotechnology, agrarian (green) biotechnology, modern (white) biotechnology, marine (blue) biotechnology, food biotechnology, and ecological biotechnology (Fig. 1.1.). In this section the perusers will comprehend the expected utilizations of biotechnology in a few fields like creation of prescriptions; diagnostics; therapeutics like monoclonal antibodies, immature microorganisms, and quality treatment; horticultural biotechnology; contamination control (bioremediation); modern and marine biotechnology; and biomaterials, as well as the moral and security issues related with a portion of the items.

Clinical Biotechnology

This area of biotechnology has numerous applications and is engaged with creation of recombinant drugs, tissue designing items, regenerative medications like undifferentiated organism and quality treatment, and a lot more biotechnology items for better human existence (Fig. 1.6). Biotechnological devices produce sanitized bio-restorative specialists on modern scales. These incorporate the two novel specialists and specialists previously accessible just in little amounts. Unrefined immunizations were utilized in ancient times in China, India, and Persia. For instance, immunization with scabs that contained the smallpox infection was a training in the East for a really long time. In 1798 English nation specialist Edward Jenner exhibited that vaccination with discharge from wounds because of disease by a connected cowpox infection could forestall smallpox undeniably less hazardously. It denoted the start of immunization. People have been benefited endlessly from the execution of inoculation programs.

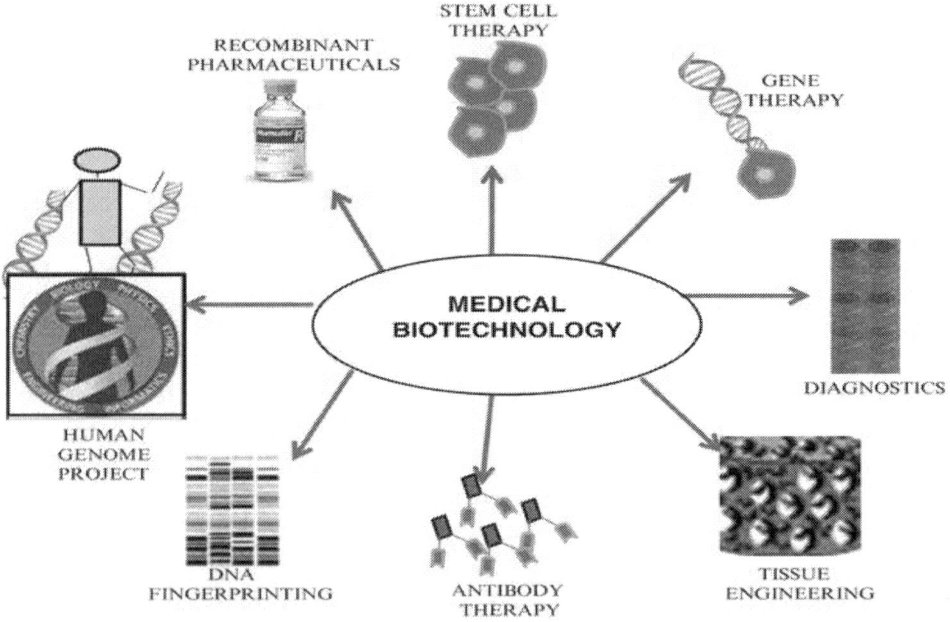

Various Applications of Medical Biotechnology

Tremendous progress has been made since the early recombinant DNA technology (RDT) experiments from which the lively—and highly profitable—biotechnology industry emerged. RDT has fomented multiple revolutions in medicine. Safe and improved drugs, accelerated drug revelation, better symptomatic and speedy strategies for identifying a disease either dynamic or dormant, advancement of new and safe immunizations, and totally clever classes of therapeutics and other clinical applications are added credits to its. The innovation has upset comprehension of infections as different as cystic fibrosis and malignant growth. Drug biotechnology being one of the significant areas includes utilizing creatures or crossovers of cancer cells or leukocytes or cells (prokaryotic, mammalian) to deliver more secure, more adequate, and financially savvy renditions of expectedly created biopharmaceuticals.

The send off of the new biopharmaceutical or drug requires screening and advancement. Mice were generally utilized as examination creatures for screening. Notwithstanding, right after creature assurance, creature cell culture offers exact and economical wellspring of cells for drug screening and viability testing. Drug biotechnology's most prominent expected lies in quality treatment and undeveloped cell based treatment. The basic reason for deformity of many acquired infections is presently found and portrayed. Quality treatment is the addition of the utilitarian quality instead of inadequate quality into cells to forestall, control, or fix infection. It likewise includes expansion of qualities for favorable to medication or cytokines to dispose of or stifle the development of the growths in disease treatment.

Be that as it may, the advancement so far is seen by numerous researchers as just a start. That's what they trust, not long from now, the refinement of "designated treatments" ought to decisively further develop drug security and viability. The improvement of prescient advancements might prompt another period in illness counteraction, especially in a portion of the world's quickly creating economies. However the dangers can't be disregarded as new turns of events and disclosures suggest new conversation starters, especially in regions as quality treatment, the morals of undifferentiated cell research, and the abuse of genomic data.

Numerous bio-remedial specialists in clinical use are biotech drugs. Insulin was among the earliest recombinant medications. Canadian physiologists Frederick Banting and Charles Best found and confined insulin in 1921. In that time numerous patients determined to have diabetes passed on inside a couple of years. During the 1960s, a few gatherings detailed integrating the chemical.

The first "bioengineered" drug, a recombinant type of human insulin, was endorsed by the US Food and Medication Organization (FDA) in 1982. Up to that point, insulin was gotten from a restricted stockpile of meat or pork pancreas tissue. By embedding the human quality for insulin into microbes, researchers had the option to accomplish lifesaving insulinproduction in enormous amounts. Sooner rather than later, patients with diabetes might have the option to breathe in insulin, killing the requirement for infusions. Breathed in insulinproducts like

Exubera® were supported by the USFDA; be that as it may, it was pulled out and different items like Technosphere® insulin (Afrezza®) are being scrutinized. They might give alleviation from prandial insulin. Afrezza comprises of a pre-dinner insulinpowder stacked into a cartridge for oral inward breath.

Technosphere innovation: The innovation permits organization of therapeutics through pneumonic course which in any case were expected to be given as infusions. These details have expansive range of physicochemical qualities and are ready with a different combination of medications with protein or little particle which might be hydrobhobic or hydrophilic or anionic or cationic in nature. The innovation can have its relevance through pneumonic course as well as for different courses of organization including nearby lung conveyance.

The principal recombinant immunization, endorsed in 1986, was created by cloning a quality piece from the hepatitis B infection into yeast (Merck›s Recombivax HB). The part was deciphered by the yeast›s hereditary apparatus into an antigenic protein. This was available on the outer layer of the infection that animates the resistant reaction. This stayed away from the need to extricate the antigen from the serum of individuals tainted with hepatitis B.

The Food and Medication organization (FDA) supported more biotech drugs in 1997 than in the past quite a long while consolidated. The FDA has supported numerous recombinant medications for human ailments. These incorporate Guides, sickliness, tumors (Kaposi›s sarcoma, leukemia, and colorectal, kidney, and ovarian malignant growths), certain circulatory issues, certain inherited problems (cystic fibrosis, familial hypercholesterolemia, Gaucher›s infection, hemophilia A, serious joined immunodeficiency illness, and Turner›s disorder), diabetic foot ulcers, diphtheria, genital moles, hepatitis B, hepatitis C, human development chemical inadequacy, and various sclerosis. Today there are in excess of 100 recombinant medications and antibodies. As a result of their proficiency, security, and somewhat minimal expense, sub-atomic symptomatic tests and recombinant immunizations might have specific significance for battling well established sicknesses of non-industrial nations, including leishmaniasis (a tropical disease causing fever and injuries) and jungle fever.

Undifferentiated cell research is extremely encouraging and holds huge potential to treat neurodegenerative problems, spinal string wounds, and different circumstances connected with organ or tissue misfortune.

DNA investigation is another strong procedure which analyzes DNA design [14] in the wake of performing RFLP and examining it by minisatellite rehash succession between at least two people. Its change, DNA profiling (cycle of matching the DNA profiles for STS markers in at least two people; see section 18), which uses multiflorus PCR examination of DNA of suspect and casualties, is exceptionally strong and is helpful in criminal examination, paternity debates, thus numerous other lawful issues. The examination is extremely helpful in criminal examinations and includes assessment of DNA from tests of the hair, body liquids, or skin at a crime location and correlation of these with those got from the suspects.

Worked on Indicative and Helpful Abilities

The sequencing of the human genome in 2003, has given specialists a verifiably rich "parts list" with which to all the more probable grasp the motivation behind why and how ailment happens. It has given added ability to quality enunciation profiling, a method for noticing verbalization of thousands of characteristics meanwhile on a glass slide called a microarray. This system can expect the forcefulness of illness.

The improvement of monoclonal antibodies in 1975 provoked a clinical change. The body routinely conveys countless antibodies—the immune structure proteins—that safeguard our body and discard microorganisms and other new intruders. By consolidating neutralizer making cells with myeloma cells, specialists had the choice to deliver antibodies that would, like "magic slugs," attach with express targets including stand-out markers, called antigenic determinants (epitopes), on the surfaces of combustible cells. When marked with radioisotopes or other separation subject matter experts, monoclonal antibodies can help in recognizing the area of dangerous development cells, thusly dealing with the precision of operation and radiation therapy and showing up—inside 48 h—whether a development is noting chemotherapy.

The polymerase chain response, a technique for enhancing smidgens of DNA previously portrayed during the 1980s, has been vital to the improvement of blood tests that can rapidly decide openness to the human immunodeficiency infection (HIV). Hereditary testing as of now is accessible for the majority intriguing monogenic problems, like hemophilia, Duchenne strong dystrophy, sickle cell weakness, thalassemia, and so on.

Another quickly creating field is proteomics, which manages investigation and ID of proteins. The investigation is finished by two-layered gel electrophoresis of the example and afterward performing mass spectrometric examination for every individual protein. The method might be useful in distinguishing the sickness related protein in the organic example. They might demonstrate early indications of infection, even before side effects show up. One such marker is C-receptive protein, a sign of provocative changes in vein walls that augur atherosclerosis.

Nanomedicine is a quickly moving field. Researchers are fostering a wide assortment of nanoparticles and nanodevices, barely a millionth of an inch in measurement, to further develop recognition of disease, support resistant reactions, fix harmed tissue, and obstruct atherosclerosis. The FDA has supported a paclitaxel egg whites settled nanoparticle detailing (Abraxane® for injectable suspension, made by Abraxis Bio Science) for the treatment of metastatic adenocarcinoma of the pancreas. Nanoparticles are being investigated in heart patients in the USA as a method for keeping their heart veins open following angioplasty.

Remedial proteins are those, which can supplant the patients normally happening proteins, when levels of the regular proteins are low or missing because of the sickness. High-throughput screening, directed with complex mechanical and PC advances, empowers researchers to test a huge number of little particles in a solitary day for their capacity to tie to or tweak the action of a «target,» like a receptor for a synapse in the cerebrum. The objective is to work on the speed and exactness of restorative protein or potential medication disclosure while bringing down the expense and working on the wellbeing of drugs that come to advertise.

A significant number of the particles used for location likewise have remedial potential as well, for instance, monoclonal antibodies. The monoclonal antibodies are supported for the therapy of numerous sicknesses as malignant growth, different sclerosis, and rheumatoid joint inflammation. They are as of now being tried in patients as possible medicines for asthma, Crohn›s illness, and strong dystrophy. As the antibodies might be effectively designated against a specific cell surface marker, in this manner they are utilized to convey a deadly portion of poisonous medication to malignant growth cells, staying away from blow-back to local typical tissues.

Rural Biotechnology

The manhas gained gigantic headway in crop improvement regarding yield; still a definitive creation of harvest is not exactly their full hereditary potential. There are many reasons like natural anxieties (atmospheric condition as downpour, cold, ice), sicknesses, bugs, and numerous different elements which diminish a definitive wanted yield influencing crop efficiency. The endeavors are proceeding to configuration crops which might be become regardless of their season or can be filled in ice or dry spell conditions for most extreme use of land, which would eventually influence crop efficiency [24]. Horticultural biotechnology means to present economical farming practices with best yield potential and insignificant unfriendly consequences for climate (Fig. 1.7). For instance, fighting vermin was a significant test. In this way, the quality from bacterium, the Tb quality, that capabilities as bug safe quality when embedded into crop plants like cotton, corn, and soybean forestalls the attack of microbe, and the device is called. This administration is useful in decreasing utilization of possibly risky pesticides on the yield. Not just the insignificant or low use of pesticides is gainful for the yield yet in addition the heap of the contaminating pesticides on climate is extraordinarily diminished.

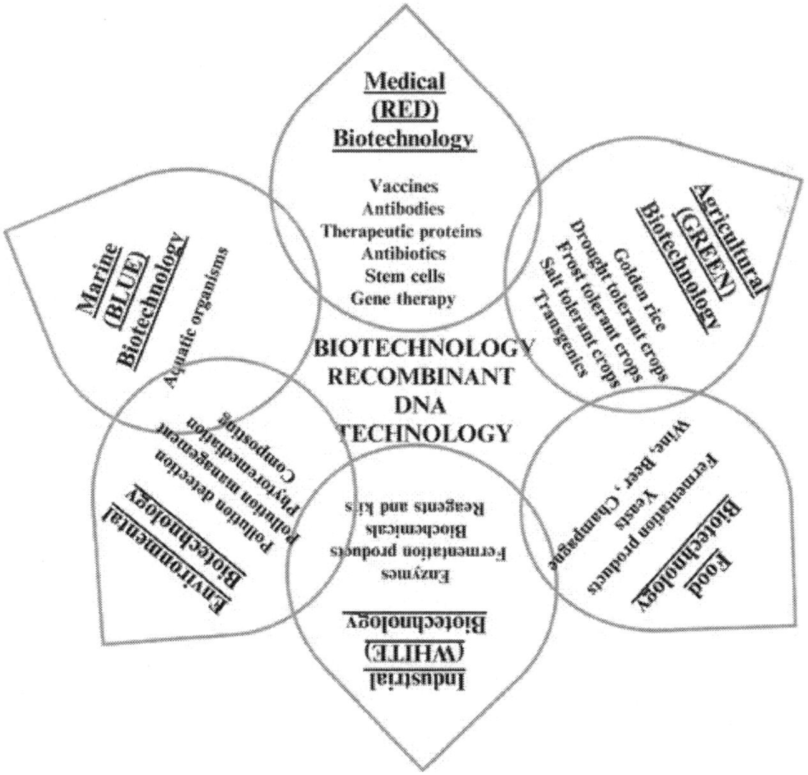

Major applications of biotechnology in different areas and some of their important products

The biotechnology appeared hundreds of years prior when plants and creatures started to be specifically reproduced and microorganisms were utilized to make brew, wine, cheddar, and bread. Nonetheless, the field step by step advanced, and as of now it is the utilization or control of living creatures to deliver recipient substances which might have clinical, horticultural, and additionally modern usage. Customary biotechnology is alluded to as the strategy that utilizes residing living being for explicit purposes as bread/cheddar making, while present day biotechnology manages the procedure that utilizes cell particles like DNA, monoclonal antibodies, biologics, and so on. Before we go into specialized advances of DNA and consequently recombinant DNA innovation, let us have the essential figuring out about DNA and its capability.

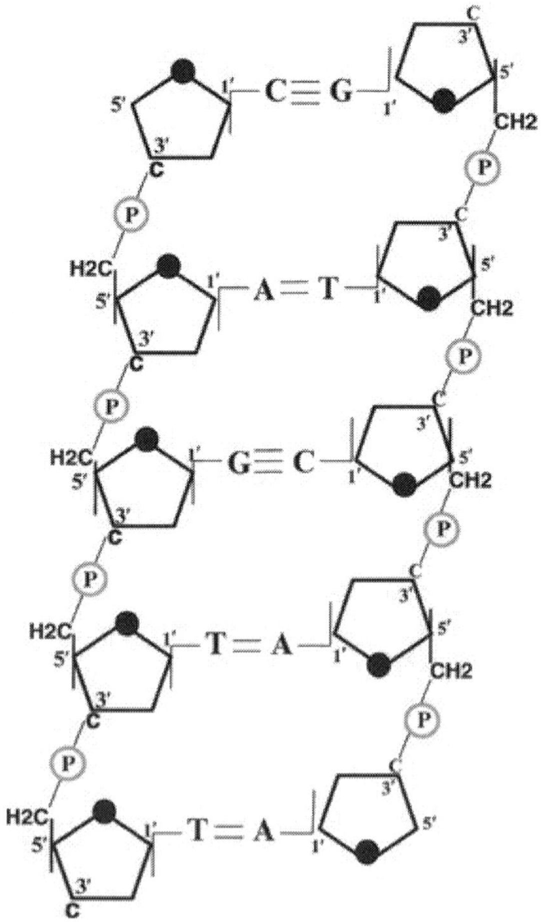

The groundwork of biotechnology was set down after the revelation of design of DNA in the mid 1950s. The innate material is deoxyribonucleic corrosive (DNA) which contains all the data that directs every single step of a singular's life. The DNA comprises of deoxyribose sugar, phosphate, and four nitrogenous bases (adenine, guanine, cytosine, and thymine). The base and sugar on the whole structure nucleoside, while base, sugar, and phosphate structure nucleotide. These are organized specifically direction on DNA called request or succession and contain data to communicate them as protein. DNA has twofold helical design, with two strands being free and antiparallel to one another, in which an on one strand base matches with T and G base matches with C with two and three bonds, separately. DNA is the long yet minimized atom which is well bundled in our core. The DNA is fit for making more duplicates such as itself with the data present in it, as request or grouping of bases. This is called DNA replication. At the point when the cell partitions into two, the DNA likewise repeats and partitions similarly into two. The course of DNA replication is displayed featuring significant stages.

The twofold helical design of DNA where the two strands are running in inverse bearing. Prolongation of the chain happens because of arrangement of phosphodiester connection between phosphate at 5′ and hydroxyl gathering of sugar at 3′ of the neighboring sugar of the nucleotide in 5-3′ bearing. The sugar is connected to the base. Bases are of four sorts: adenine (A), guanine (G) (purines), thymine (T), and cytosine (C) (pyrimidines). Adenine base matches with two hydrogen bonds with thymine on the inverse antiparallel strand and guanine base matches with three hydrogen bonds with cytosine present on the inverse antiparallel strand.

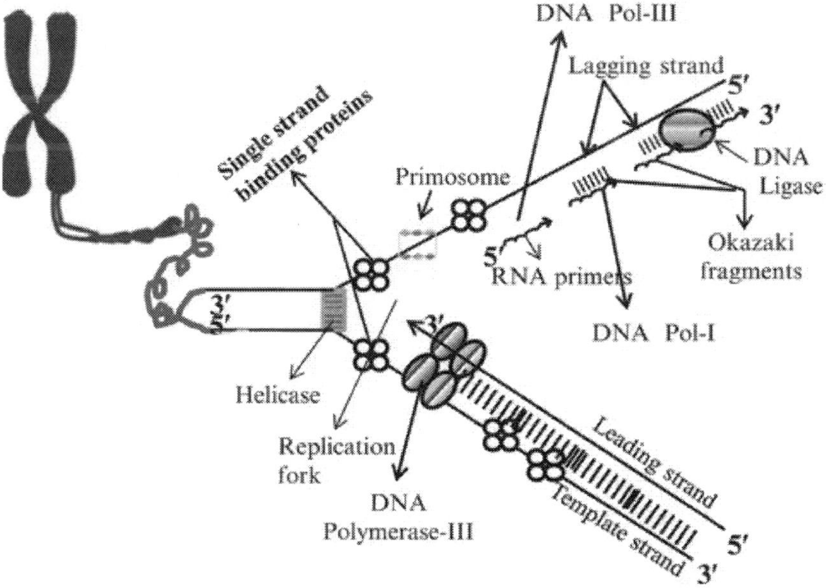

The course of DNA replication. The DNA is thickly pressed and bundled in the chromosomes. The interaction requires the activity of a few variables and proteins. DNA helicase loosens up the twofold helix. Topoisomerase loosens up DNA from its really wound nature. Single-strand restricting proteins tie to single-abandoned open DNA and forestall its reannealing and keeps strand distance. DNA polymerase is a chemical which fabricates another free DNA strand and has editing action. DNA brace is a protein which forestalls separation of DNA polymerase. Primase gives a short RNA grouping to DNA polymerase to start blend. DNA ligase reanneals and joins the Okazaki sections of the slacking strand. DNA duplication follows semiconservative replication, where each strand fills in as layout which prompts the development of two free strands. In the recently shaped DNA, one strand is old and the other one is new (semiconservative replication). DNA polymerase can broaden existing short DNA or RNA strand which is matched to layout strand and is called preliminary. Preliminary is expected as DNA polymerase can't begin the combination straightforwardly. DNA polymerase is equipped for editing, or at least, amendment of wrongly consolidated nucleotide. One strand is recreated ceaselessly with single preliminary, and it is called as driving strand. Other strand is intermittent and requires the expansion of a few preliminaries. The expansion is finished as short parts called as Okazaki pieces. The holes are fixed by DNA ligase. Replication generally happens in 5'-3' bearing.

DNA contains entire data for the working of the cell. The piece of the DNA which has data to direct the biosynthesis of a polypeptide is known as a "quality." The plan or request of nucleotides decides the sort of proteins which we produce. Every quality is liable for coding a useful polypeptide. The qualities have the data to make a free duplicate of mRNA. The data of DNA which makes RNA thus assists cells with consolidating amino acids as indicated by course of action of letters for making numerous sorts of proteins. These letters are deciphered into mRNA as trio codon, where every codon indicates one specific amino corrosive. The polypeptide is consequently made by adding particular amino acids as per the guidelines present on RNA. Accordingly, the game plan of four bases (adenine, guanine, cytosine, and thymine) directs the data to add any of the 20 amino acids to make every one of the proteins in every one of the living creatures. Scarcely any qualities should be communicated consistently, as their items are expected by the cell, and these are known as "constitutive qualities." They are answerable for essential housekeeping elements of the cells. Nonetheless, contingent on the physiological interest and cell's prerequisite at a specific time, a few qualities are dynamic and some are latent, or at least, they code for no protein. The data contained in the DNA is utilized to make mRNA during the time spent " record" (factors displayed in Table 1.1). The data of mRNA is utilized during the time spent " interpretation" for creation of protein. Record happens in the core and interpretation in the cytoplasm of the cell. In interpretation a few commencement factors assist in the gathering of mRNA with 40S ribosome and forestall restricting of both ribosomal subunits; they likewise partner with cap and poly(A) tail. A few extension factors assume a significant part in chain prolongation. However every cell of the body has a similar hereditary cosmetics, yet each is particular to carry out exceptional roles, the enactment and articulation of qualities is different in every cell. Consequently, one sort of cells can communicate a portion of its qualities all at once and may not communicate similar qualities later. This is designated "worldly guideline" of the quality. In the body various cells express various qualities and subsequently various proteins. For instance, liver cell and lymphocyte, would communicate various qualities. This is known as spatial guideline of the quality. Hence, in the cells of the body, the actuation of qualities is under spatial guideline (cells present at various areas and various organs produce various proteins) and worldly guideline (same cells produce various proteins at various times). The proteins are framed by the data contained in the DNA and play out various cell capabilities. The proteins might be underlying (answerable for cell shape and size), or they might be utilitarian like catalysts, flagging intermediates, administrative proteins, and protection framework proteins. Nonetheless, any sort of hereditary imperfection brings about deficient protein or modifies protein collapsing which can think twice about working of the body and is liable for the sicknesses. Figure 1.4 shows the blueprint of the course of record and interpretation with significant stages.

Definition

Biotechnology- Bio means life and technology means the application of knowledge for practical use *ie.*, the use of living organisms to make or improve a product.

Other definitions for the term Biotechnology

- The use of living organisms to solve problems or make useful products.

- The use of cells and biological molecules to solve problems or make useful products. Biological molecules include DNA, RNA and proteins.
- The commercial application of living organisms or their products, which involves the deliberate manipulation of their DNA molecules.
- Make a living cell to perform a specific task in a predictable and controllable way.

Biotechnology has been described as "Janus-faced". This implies that there are two sides. On one side techniques allow DNA to be manipulated to move genes from one organism to another. On the other, it involves relatively new technologies whose consequences are untested and should be met with caution.

Stages of biotechnology development

- Ancient biotechnology - 8000-4000 B.C

  Early history as related to food and shelter; includes domestication
- Classical biotechnology - 2000 B.C.; 1800-1900 AD

  Built on ancient biotechnology; fermentation promoted food production and medicine
- 1900-1953: Genetics
- 1953 - 1976: DNA research, science explodes
- Modern biotechnology - 1977

  Manipulates genetic information in organism; Genetic engineering

Biotechnology is a collection of various technologies that enable us to improve crop yield and food quality in agriculture and to produce a broader array of products in industries.

Various technologies and their uses

- Genetic Engineering (Recombinant DNA) Technology

  The use of cellular enzymes to manipulate DNA

  Transferring DNA between unrelated organisms
- Protein Engineering Technology

  Improve existing/create novel proteins to make useful products
- Antisense or RNAi Technology

  Block or decrease the production of certain proteins
- Cell and Tissue Culture Technology

  Grow cells/tissues under laboratory conditions to produce an entire organism, or to produce new products

- Bioinformatics Technology

  Computational analysis of biological data, e.g., sequence analysis macromolecular structures, high-throughput profiling data analysis

- Functional Genomics (the -omics)

  The use of genome-wide, high-throughput approaches to determine the biological function of all of the genes and their products

High-throughput technologies (the-omics)

Transcriptomics (for example microarray articulation profiling)

Proteomics (for example structures/alterations/connections of proteins)

Proteins are liable for a perpetual number of errands inside the cell. The total arrangement of proteins in a cell can be alluded to as its proteome and the investigation of protein design and capability and what each protein in the cell is doing is known as proteomics. The proteome is exceptionally unique and it changes now and again because of various ecological upgrades. The objective of proteomics is to comprehend how the construction and capability of proteins permit them to do what they do, what they connect with and how they add to life processes.

Metabolomics (for example metabolite profiling, substance fingerprinting, transition examination)

Metabolomics is one of the freshest 'omics' sciences. The metabolome alludes to the total arrangement of low sub-atomic weight intensifies in an example. These mixtures are the substrates and results of enzymatic responses and straightforwardly affect the aggregate of the cell. Subsequently, metabolomics targets deciding an example's profile of these mixtures at a predefined time under unambiguous natural circumstances.

Trans genomics (for example take out, thump in, quality labeling, mutagenesis) • Translational genomics.

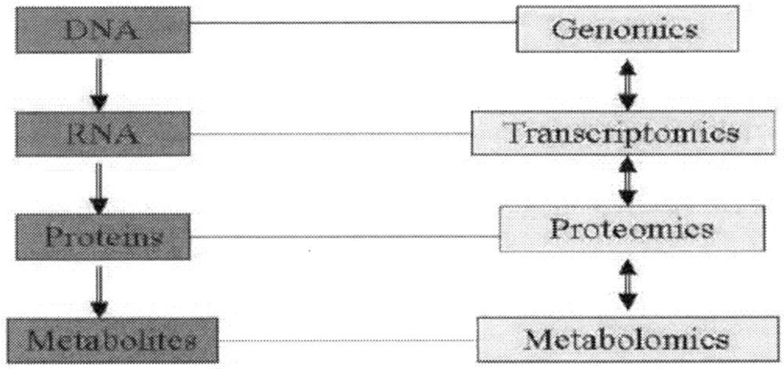

Applications of biotechnology & genomics 1. Environmental biotechnology

A. Environmental monitoring

- Diagnosis of environmental problems via biotechnology

B. Waste management

- Bioremediation: the use of microbes to break down organic molecules or environmental pollutants.
- Phyto remediation: the use of plants to remove pollutants (e.g. heavy metals) from the environment.

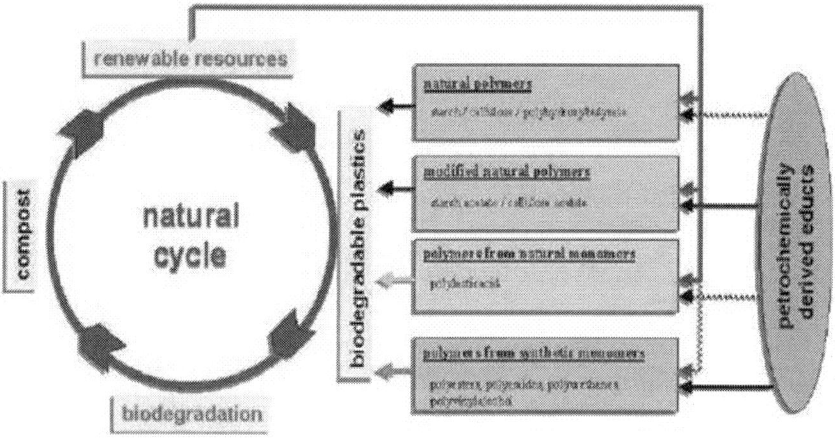

C. Pollution prevention

- Renewable resources
- Biodegradable products
- Alternative energy sources

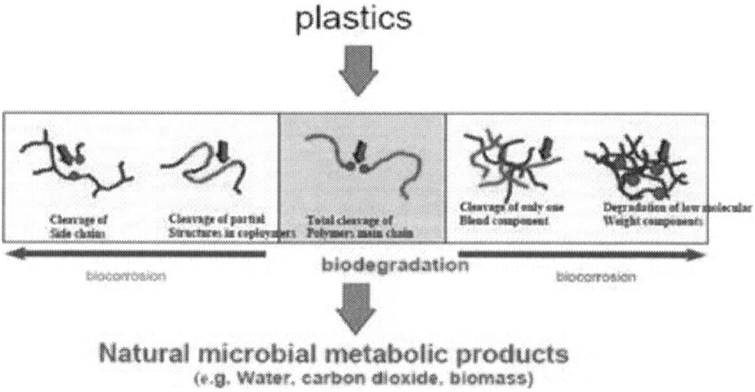

2. Medical biotechnology

A. Diagnostics

B. Therapeutics

C. Vaccines

D. Medical research tools

E. Human Genome Research

3. Agricultural biotechnology

A. Animal Biotechnology

B. Crop Biotechnology

C. Horticultural Biotechnology

D. Tree Biotechnology

E. Food processing

4. Evolutionary and ecological genomics

Finding genes associated with ecological traits and evolutionary diversification. Common goals: health, productivity

Plant biotechnology/Agricultural biotechnology

A process to produce a genetically modified plant by removing genetic information from an organism, manipulating it in the laboratory and then transferring it into a plant to change certain of its characteristics. In Nutshell it's the manipulation of plants for the benefit of mankind

The plants are mainly manipulated for two major objectives

A. Crop improvement

- Herbicide tolerance (in use)
- Pest resistance (in use)
- Drought tolerance
- Nitrogen fixing ability
- Acidity and Salinity tolerance

B. Nutritional value of crops

- Improving food quality and safety
- Healthier cooking oils by decreasing the conc. of saturated fatty acids in vegetable oils
- Functional foods: foods containing significant levels of biologically active components that impart health benefits

Various technologies applied in plant biotechnology includes

- Genetic engineering/ recombinant DNA technology
- Tissue culture
- Molecular breeding - MAS

Traditional plant breeding involves cross-breeding of similar plants to produce new varieties with different traits. But it takes many generations to achieve desired result. By using various biotechnological tools, crop improvement can be achieved faster and it even facilitates to transfer genes from unrelated species.

Genetic Engineering

Manipulation of genes is called genetic engineering or recombinant DNA technology. It removes gene(s) from one organism and either

- Transfers them to another
- Puts them back in the original with a different combination

Various gene transfer techniques used in genetic engineering includes

- Agrobacterium mediated gene transfer: Desired trait is isolated from DNA of original organism, inserted into Agrobacterium, target plant is infected. Cells that accept the DNA are grown into plants with the new trait.
- Gene gun: DNA that codes for the desired trait is coated onto tiny particles of tungsten and fired into a group of plant cells. Cells that accept the DNA are grown into plants with the desired trait.

Tissue Culture

Tissue culture manipulates cells, anthers, pollen grains, or other tissues; so they live for extended periods under laboratory conditions or become whole, living, growing organisms; genetically engineered cells may be converted into genetically engineered organisms through tissue culture.

Marker Assisted Selection

Marker-aided genetic analysis studies DNA sequences to identify genes, QTLs (quantitative trait loci), and other molecular markers and to associate them with organism functions, i.e., gene identification. Marker-aided selection is the identification and inheritance tracing of previously identified DNA fragments through a series of generations.

Applications of biotechnology in agriculture (plants)

A. Crop Improvement

- Plants with built in resistance to pest and Diseases.
- Plants with built in tolerance to environmental conditions
- Improved color and quality

B. Pharmaceuticals

- Plants that produce edible vaccines

C. Food

- Improved taste and nutrition
- Improved handling qualities

D. Industrial

- plants that produce plastics, fuels, and other products
- plants for environmental cleanup

E. Other

- pesticides made from naturally-occurring microorganisms and insects

Applications of biotechnology in agriculture (animals) A. Food

- Increased milk production

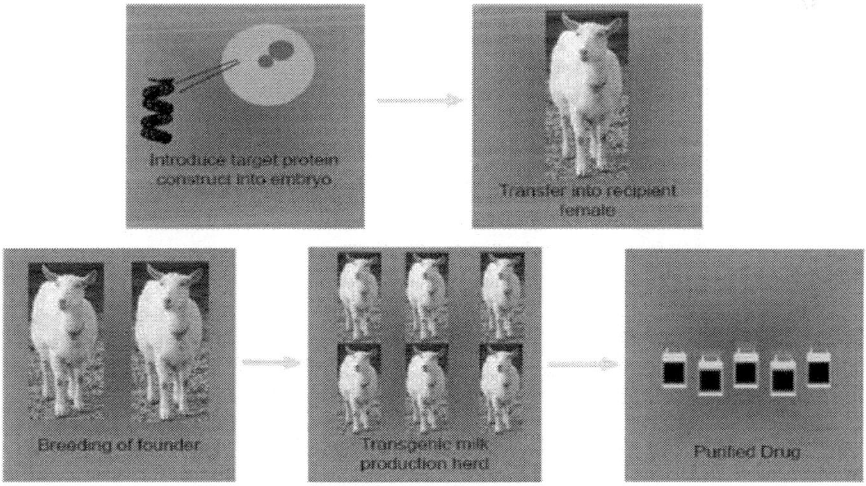

- leaner meat in pork
- growth hormones in farm-raised fish that result in earlier market-ready fish

B. Pharmaceuticals

- Animals engineered to produce human proteins for drugs, including insulin and vaccines

C. Breeding

- Disease tolerance
- Exact copies of desired stock
- Increased yields

D. Health

- Microorganisms introduced into feed for beneficial purposes
- Diagnostics for disease and pregnancy detection
- Animals engineered to produce organs suitable for transplantation into humans.

History of Biotechnology

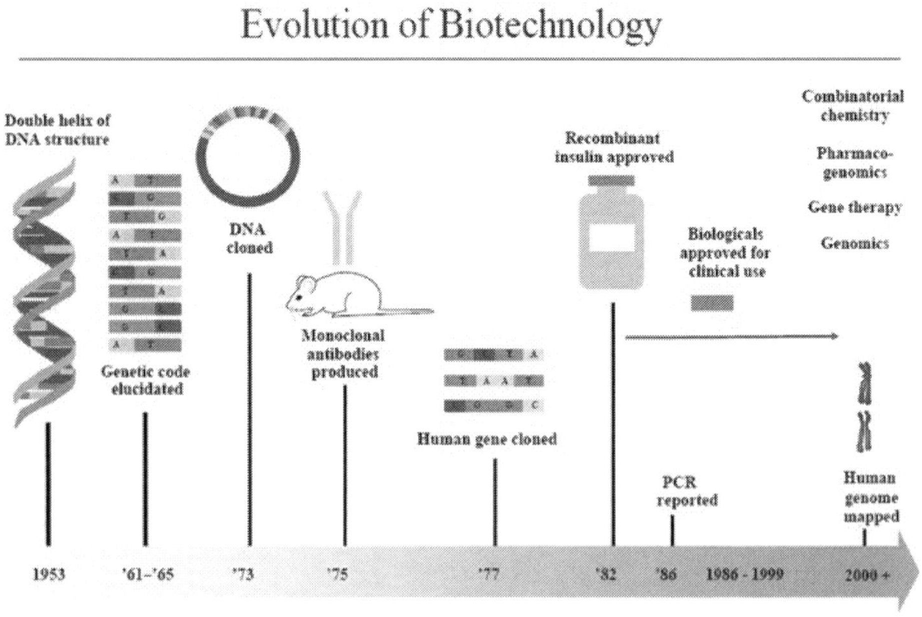

1797: First vaccination. Edward Jenner takes pus from a cowpox lesion, inserts it into an incision on a boy's arm. 1830: Proteins are discovered.

1833: First enzyme is discovered and isolated

1865: Gregor Mendel discovers the laws of inheritance by studying flowers in his garden. The science of genetics begins.

1915: Phages – viruses that only infect bacteria – are discovered

1927: Herman Muller discovers that radiation causes defects in chromosomes.

1944: DNA is proven to carry genetic information by Oswald Avery, Colin MacLeod and Maclin McCarty.

1953: James Watson and Francis Crick describe the double helical structure of DNA. They shared the 1962 Nobel Prize in Medicine or Physiology with Maurice Wilkins.

1955: The amino acid sequence of insulin is discovered by Frederick Sanger.

1958: DNA is made in a test tube for the first time. Sickle cell disease is shown to occur due to a change in one amino acid

1971: The first complete synthesis of a gene occurs. Discovery of restriction enzymes that cut and splice genetic material very specifically occurs. This opens the way for gene cloning.

1973: Stanley Cohen and Herbert Boyer perfect genetic engineering techniques to cut and paste DNA using restriction enzymes.

1975: Georges Kohler and Cesar Milstein develop the technology to produce monoclonal antibodies—highly specific, purified antibodies derived from only one clone of cells that recognize only one antigen. They shared the

1984: Nobel Prize in Physiology or Medicine with Neils Jerne.

1981: The first transgenic animals are produced by transferring genes from other animals into mice.

1983: The polymerase chain reaction (PCR) technique, which makes unlimited copies of genes and gene fragments, is conceived. Kary Mullis, who was born in Lenoir, N.C., wins the 1993 Nobel Prize in Chemistry for the discovery.

1986: First recombinant vaccine is approved for human use: hepatitis B. First anti-cancer drug is produced through biotech: interferon.

1987: First approval for field tests of a genetically modified food plant: virus-resistant tomatoes.

1994: Genetically modified tomatoes are sold in the U.S. for the first time.

1990: The Human Genome Project – an international effort to maps all of the genes in the human genome – is launched.

2002: The draft version of the human genome is published.

1997: Scientists report the birth of Dolly, the first animal cloned from an adult cell. 1998:Human embryonic stem cell lines are established. They offer hope to many because they may be able to replace diseased or dysfunctional cells.

2003: The SARS (severe acute respiratory syndrome) virus is sequenced three weeks after its discovery.

2004: The first cloned pet – a kitten – is delivered to its owner. She is called CopyCat (or Cc for short).

2006: A recombinant vaccine against human papillomavirus (HPV) receives FDA approval.

The virus causes genital warts and can cause cervical cancer.

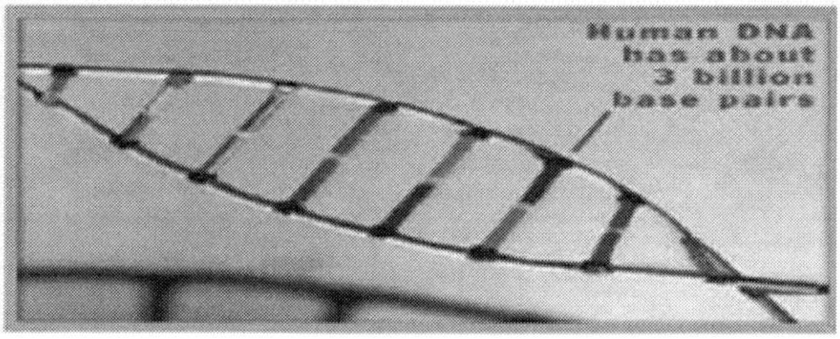

1. The term "Biotechnology" was first coined in ....
   a) **1919** b) 1916
   c) 1991 d) 1961
2. The term "Biotechnology" was first coined by ....
   a) Neuberg b) Lipmann
   c) Frederick Sanger's d) **Karl Ereky**
3. Biotechnology is used in areas including ....
   a) Agriculture b) Bioremediation
   c) Food processing d) **All the above**
4. In agriculture, genetic engineering is being used to produce plants that are resistant to ....
   a) insects b) weeds
   c) plant diseases d) **All the above**

5. Immunoassays are used for ....
   a) drug level testing
   b) detection of unsafe levels of pesticides, herbicides and toxins on crops and in animal products
   c) **both a anb b**
   d) None of the above
6. Genetic engineering is ....
   a) **use of cellular enzymes to manipulate DNA**
   b) Improve existing/create novel proteins to make useful products
   c) Block or decrease the production of certain proteins
   d) Grow cells/tissues under laboratory conditions to produce an entire organism, or to produce new products
7. Protein engineering is ....
   a) use of cellular enzymes to manipulate DNA
   b) **Improve existing/create novel proteins to make useful products**
   c) Block or decrease the production of certain proteins
   d) Grow cells/tissues under laboratory conditions to produce an entire organism, or to produce new products
8. Antisense or RNAi technology is ....
   a) use of cellular enzymes to manipulate
   b) Improve existing/create novel proteins to DNA make useful products
   c) **Block or decrease the production of proteins conditions to produce an entire**
   d) Grow cells/tissues under laboratory certain organism, or to produce new products
9. Cell and tissue culture technology is ....
   a) use of cellular enzymes to manipulate
   b) Improve existing/create novel proteins to DNA make useful products
   c) Block or decrease the production of conditions to produce an entire
   d) **Grow cells/tissues under laboratory certain proteins organism, or to produce new products**

10. Bioinformatics technology is ....
    a) use of cellular enzymes to manipulate DNA
    b) Improve existing/create novel proteins to make useful products
    c) Block or decrease the production of certain proteins
    d) **Computational analysis of biological data**
11. Functional Genomics
    a) **High-throughput approaches to determine the biological function of all of the genes and their products**
    b) Improve existing/create novel proteins to make useful products
    c) Block or decrease the production of certain proteins
    d) Computational analysis of biological data
12. Translational genomics include(s)..............
    a) Transcriptomics alone
    b) Proteomics alone
    c) Metabolomics and Transgenomics
    d) **All the above**
13. Metabolomics include(s) ....
    a) Metabolite profiling   c) Flux analysis
    b) Chemical fingerprinting   d) **All the above**
14. Various technologies applied in plant biotechnology includes ....
    a) Genetic engineering/ recombinant
    b) Tissue culture DNA technology
    c) Molecular breeding - MAS
    d) **All the above**
15. Vaccination was first attempted by ....
    a) Gregor Mendel   c) Colin MacLeod
    b) Oswald Avery   d) **Edward Jenner**
16. Vaccination was first attempted in the year........................
    a) **1797**   b) 1777
    c) 1787   d) 1767

17. Laws of inheritance was discovered by ....
    a) **Gregor Mendel** b) Oswald Avery
    c) Colin MacLeod d) Edward Jenner
18. The science of genetics was born in the year......................
    a) **1865** b) 1856
    c) 1855 d) 1846
19. Phages infect ....
    a) **Only bacteria** b) Only viruse
    c) Both a & b d) None of the above
20. Name the scientist who discovered that radiation causes defects in chromosomes ....
    a) Gregor Mendel b) Oswald Avery
    c) Colin MacLeod d) **Herman Muller**
21. DNA is proven to carry genetic information by......................
    a) Oswald Avery b) Colin MacLeod
    c) Maclyn McCarty d) **All the above**
22. DNA is proven to carry genetic information in the year ....
    a) 1944 b) 1946
    c) 1948 d) None of the above
23. Double helical structure of DNA is described by ....
    a) James Watson b) Francis Crick
    c) **Both** d) None of the above
24. The amino acid sequence of insulin is discovered by ....
    a) James Watson b) Francis Crick
    c) **Frederick Sanger** d) None of the above
25. Sickle cell disease occurs due to a change in ....amino acid(s).
    a) **One** b) Two
    c) Three d) None of the above
26. The first complete synthesis of a gene occurred in the ....
    a) 1917 b) **1971**
    c) 1791 d) None of the above

27. Genetic engineering technique, to cut and paste DNA using restriction enzymes is perfected by ....

a) Stanley Cohen b) Herbert Boyer

c) **Both** d) None of the above

28. The technology to produce monoclonal antibodies is developed by ....

a) Georges Kohler b) Cesar Milstein

c) **Both** d) None of the above

29. .... developed the polymerase chain reaction (PCR) technique.

a) **Kary Mullis** b) Georges Kohler

c) Cesar Milstein d) None of the above

30. First recombinant vaccine approved for human use is ....

a) **hepatitis B** b) polio

c) hepatitis C d) None of the above

31. .... is the first genetically modified plant approved for field tests.

a) **virus-resistant tomatoes** b) virus-resistant brinjal

c) virus-resistant cotton d) None of the above

32. Genetically modified tomatoes are sold in the U.S. for the first time in the year ....

a) **1994** b) 1971

c) 1991 d) None of the above

33. The Human Genome Project is launched in ....

a) **1990** b) 1991

c) 2000 d) None of the above

34. The draft version of the human genome is published in ....

a) 1990 b) 2000

c) **2002** d) 2004

References

1. Behera et al. (2015) Scope of algae as third generation biofuels. Front Bioeng Biotechnol 2: doi:10.3389/fbioe.2014.00090 [PMC free article] [PubMed]
2. Bruhn CM. Consumer concerns and educational strategies: focus on biotechnology. Food Technol. 1992;46:80–102. [Google Scholar]
3. Council for Agricultural Science and Technology: "Applications of Biotechnology to Crops: Benefits and Risks", Issue Paper, Number 12, Dec. 1999
4. Definition of Biotechnology-Economic Research Service at United States Department of Agriculture
5. Erickson B, Nelson JE, Winters P. Perspective on opportunities in industrial biotechnology in renewable chemicals. Biotechnol J. 2012;7:176–185. doi: 10.1002/biot.201100069. [PMC free article] [PubMed] [CrossRef] [Google Scholar]
6. Food biotechnology-benefits for the developing countries. http://ificinfo.health.org/insight/janfeb99/foodbiotechnology.htm
7. Food biotechnology: health and harvest for our times. http://ificinfo.health.org/brochure/biobroch.htm
8. Genetically engineered foods, fears and facts: FDA. Consumer 27(1). January/February 1993, pp. 11–14. http://www.fda.gov/fdac/100_toc.html
9. Hoban T, Kendall P. Consumer attitudes about food biotechnology: final project report. Raleigh: North Carolina Cooperative Extension Service, North Carolina State University; 1994. [Google Scholar]
10. IFIC Foundation- Americans remain positive on food biotechnology. http://ificinfo.health.org/press/positivebio.htm
11. Jikia et al (2005) The use of a novel biodegradable preparation capable of the sustained release of bacteriophages and ciprofloxacin, in the complex treatment of multidrug-resistant Staphylococcus aureus-infected local radiation injuries caused by exposure to Sr90. Clin Exp Dermatol 30:23–6 [PubMed]
12. Myths and facts about food biotechnology, food insight, September/October 1999, pp. 2–3
13. National and international policy making in biotechnology. http://www.biotechknowledge.com/showlib.php3?194
14. New York Times Editorial, titled: "Food....people who would have the most to lose", 19 Nov 1999
15. North Carolina Biotechnology Center. "About Biotech". http://www.ncbiotech.org/aboutbt/main.cfm
16. Principles of biotechnology. http://www.nal.usda.gov/bic/Education_res/iastate.info/bio1.html
17. Public perceptions of biotechnology. A summary of research by Dr. Thomas Hoban at North Carolina State University. http://www4.ncsu.edu/~hobantj/biotech.htm
18. The Council for Biotechnology Information. http://www.whybiotech.com
19. Thieman WJ, Palladino MA (2004) Introduction to biotechnology, 2nd edn. Pearson Publications, USA

20. U.S. Food and Drug Administration Center for food Safety and Applied Nutrition: October-November 1997. http://vm.cfsan.fda.gov/~tdms/pubalgry.html
21. U.S. Food and Drug Administration Center for food Safety and Applied Nutrition: Q & A Sheet: June 1992. http://vm.cfsan.fda.gov/~lrd/bioqa.html
22. United States Department of Agriculture "Agricultural Biotechnology Concepts and Definitions". http://www.biotechknowledge.com/showlib.php3?1739
23. van Beuzekom B, Arundel A. OECD biotechnology statistics 2006. Paris: OECD; 2006. [Google Scholar]
24. What the experts say about food biotechnology http://ificinfo.health.org/foodbiotech/whatexpertssay.htm

3

Future of Plant Biotechnology

Dr. R. A. Mir
Associate Professor, Qurat Ul Ain

ABSTRACT

Current biotechnology empowers an organic entity to create an absolutely new item which the living being doesn't or can't deliver typically through the joining of the innovation of 'Hereditary designing'. Biotechnology shows its specialized benefits and new advancement possibilities in reproducing of new plants assortments with high and stable yield, great quality, as well as pressure resilience and obstruction. Probably the most common issues looked in agrarian biological systems could be addressed with the presentation of transgenic crops consolidated with characteristics for bug bother obstruction, herbicide resilience and protection from viral illnesses. Plant biotechnology has acquired significance in the new past for expanding the quality and amount of farming, agricultural, decorative plants, and in controlling the plants for worked on agronomic execution. Late advancements in the genome sequencing will have sweeping ramifications for future agribusiness. From this review, we can realize that the creating scene embraces these quick changing advances soon and bridle their uncommon potential for the future advantage of person.

Introduction

Plant biotechnology is established on the showed to impotency of plant cells, joined with the conveyance, stable reconciliation and articulation of transgenes in plant cells, the recovery of changed plants, and the Mendelian transmission of transgenes to the posterity. Man has trained plants and creatures from the crazy around quite a while back. Determined choice throughout the long term brought about crop genotypes which are reasonable for human food. Since the revelation of the laws of heredity by Mendel in 1865, controlled reproducing altered horticulture and upgraded crop yield. Devices of recombinant DNA innovation created in 1960s introduced the period of new biosciences which would reform each feature of human existence in a protected and reasonable way in the following hundred years.

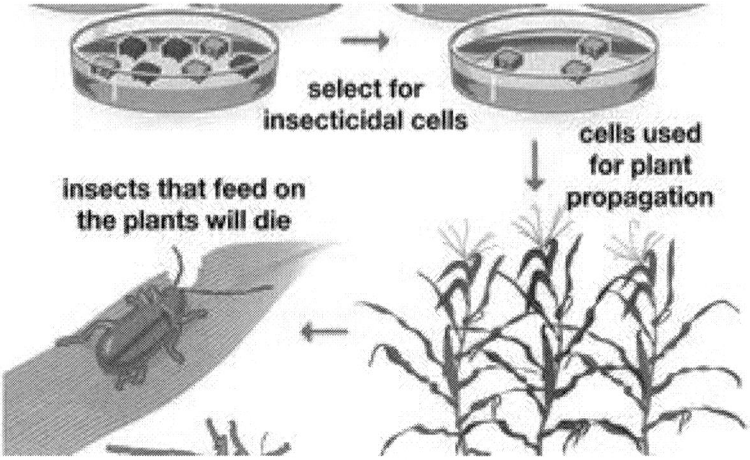

Plant biotechnology address one of various contending innovative ways to deal with resolving a specific agronomic issue yet in any case, for instance, a specific irritation issue could similarly be tended to through customary plant reproducing, through a transgenic approach, or through a coordinated crop management (ICM) approach or any mix of these. Plant biotechnology is a helpful and amazing asset for the improvement of new plant qualities and assortments. Such new assortments should be delivered for an enormous scope to make business progress and to fulfill the interest from producers. Customarily, new assortments were accomplished by the seed engendering strategy. A large number of the farming sources of info and cycles that have been ended up being destructive throughout recent many years should be gotten rid of to safeguard the biological equilibrium, ecological wellbeing and normal assets (Kumar, 2001). Biotechnology and hereditary designing of plants will assume a vital part around here.

As of now, the plantlets delivered by micro propagation offer a down to earth elective for the vast majority plant species. Momentum plant biotechnology has created as another period of science and innovation where the development of auxiliary metabolites, important plant hereditary qualities upgrades, germplasm preservation, and creation of huge quantities of infection free and new assortments are liked. This article surveys the headway made in the previous a very long time in the space of biotechnology of plants, and contemplates over the possibilities in the new thousand years.

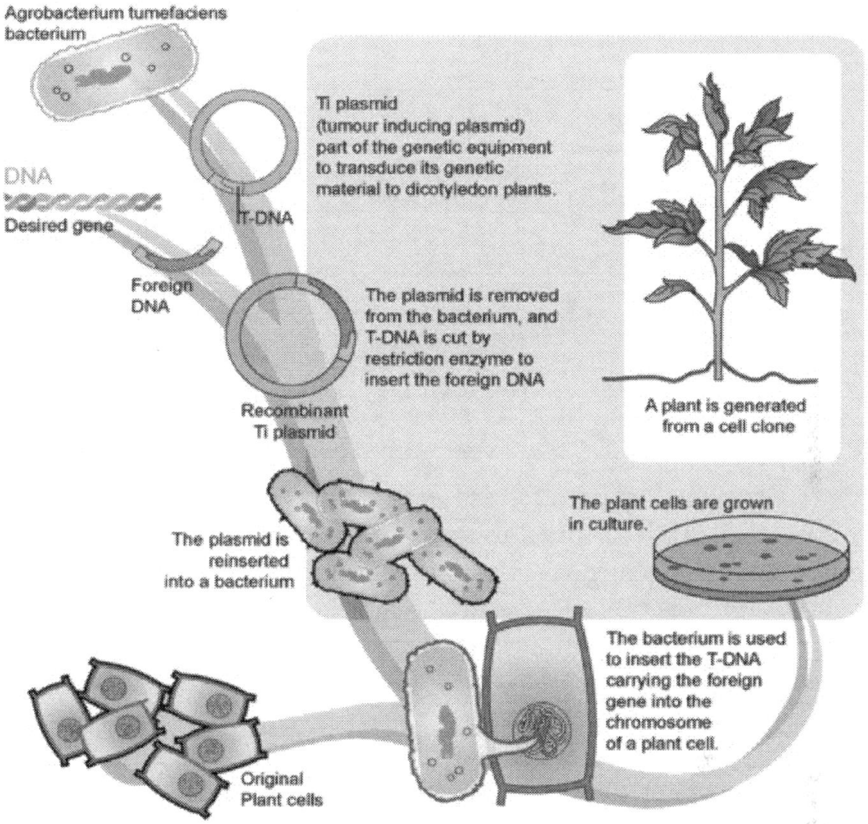

Plant Genetic Engineering

Acquaintance of qualities into plants with make new economically helpful assortments might appear to be an important errand today. In the mid 1980s in any case, this was one of the significant bottlenecks forestalling the satisfaction of a rural transformation that started following the revelation and utilization of limitation compounds, followed quickly by the hereditary designing of microbes for modern and clinical applications. Plant biotechnology has been innovation driven since its presentation, and the effective foundation of quality exchange advances for significant yields was a significant forward leap for the little biotechnology organizations that lead to improvements in the field in the mid-1980s (Gordon-kam et al., 1990). At the point when the dirt bacterium Agrobacterium tumefaction's was displayed to move part of the DNA from an inhabitant plasmid into the plant genome, it didn't take long to create the principal model transgenic plants (Barton and Binns, 1983).

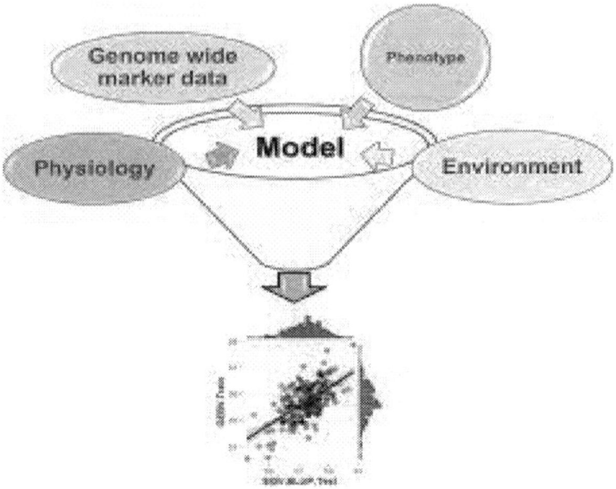

The early trailblazers of hereditary designing in plants predicted the capability of the innovation and its capacity to build yields and address our most difficult social issues, like neediness and food frailty. While the innovation has advanced quickly, the positive effect it could have all around the world is unnecessarily squandered

Biotic Stress Resistance

Biotic burdens are the harm to plants brought about by other living organic entities like microbes, parasites, nematodes, protists, bugs, infections and viroids. Various biotic anxieties are of verifiable importance, for example, the potato scourge in Ireland, espresso rust in Brazil, maize leaf curse brought about by Cochliobolus heterotrophs in the US and the incomparable Bengal starvation in 1943 (Hussain, 2015). Microorganisms represent around 15% misfortunes in worldwide food creation, and are a significant test in reproducing safe yields (Onaga and Wyre, 2016).

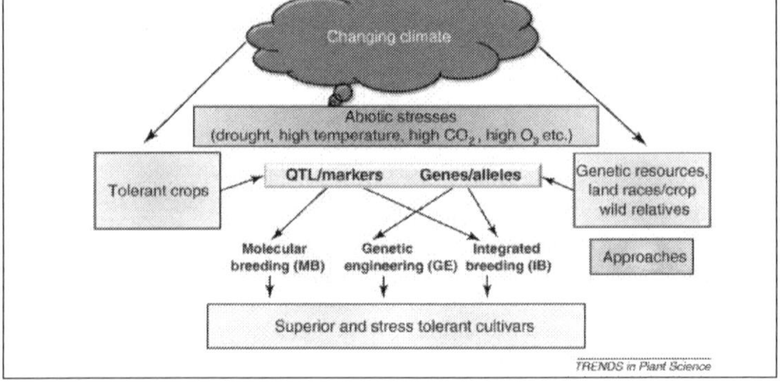

It is assessed that sickness or bug bother episodes are supposed to keep on causing food creation misfortunes or even deteriorate by growing to the areas they were not predominant previously (Siddra, 2012). For the control of this pressure, broad and regularly, aimless utilization of pesticides throughout recent many years affects human wellbeing and corruption of climate and delicate biological systems. Thus, there is a critical need to diminish the utilization of pesticides by executing the devices of biotechnology and make the yields innately impervious to irritations and sicknesses.

Insect Pest Resistance

It is assessed that 14% of yield efficiency is lost to bug bothers on a worldwide scale (Rattier, 1996). Bug safe transgenic crops were first popularized during the 1990s with the presentation of GM corn (maize), potato and cotton plants communicating qualities encoding the entomocidal δ-endotoxin from Bacillus thuringiensis Bt; otherwise called Cry proteins (Gatehouse et al., 2011). Presentation of qualities encoding insecticidal proteins of Bacillus thuringiensisinto crop plants gave stable obstruction. Escalated endeavors during the 1990s have prompted the commercialization of bug safe cotton, corn and potato (de Maagd et al., 1999).

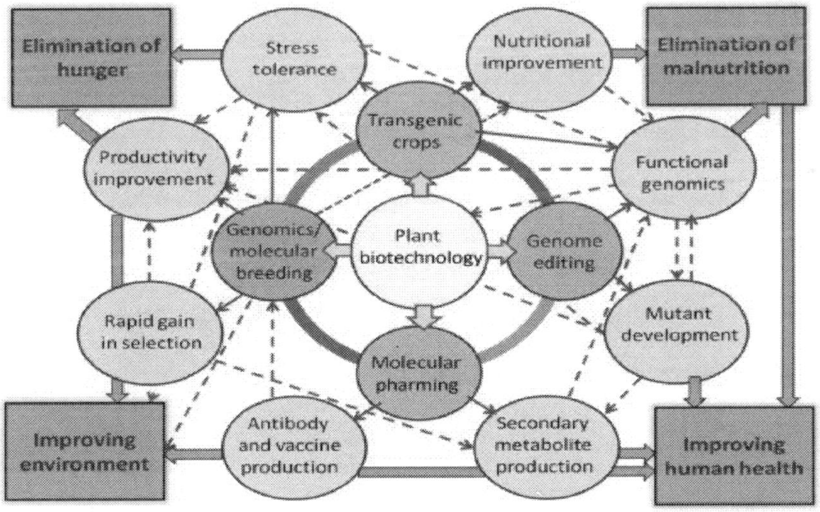

Protection from these bugs can be effectively designed by taking advantage of insecticidal proteins tracked down in microorganisms, plants and creatures (Kumar, 2001). Numerous insecticidal proteins are accessible in nature which are profoundly well defined for agronomic ally significant bug bothers and yet these are innocuous to man, warm blooded creatures and different organic entities including helpful bugs (Adamczyk and Hardee, 2002). The bacterium B. thuringiensis (Bt) was first found in Japan in 1902 in a silkworm raising unit. In 1911, it was again disengaged in a flour moth populace and described by Berliner in Thüringen (Germany). Most Bt strains produce a few glasslike proteins (Cry proteins), every one of which shows a somewhat restricted have range (Bravo et al., 2013). The cutting edge transgenic harvests to be created before long will convey different insecticidal protein qualities.

Viral Resistance

Viral illnesses make impressive misfortunes crop efficiency. Means crops, that incorporates cassava, yam, potato, banana, papaya, normal bean, rice and maize are frequently tainted with RNA as well as DNA infections that can't be controlled with pesticides (Kreuze and Valkonen, 2017). Generally speaking, opposition sources are missing, or the hereditary intricacy and hardships in introgression obstruction qualities to cultivars by crossing hamper the endeavors of yield improvement. Subsequently, improvement and move of protection from crops by biotechnological implies offers an alluring elective arrangement.

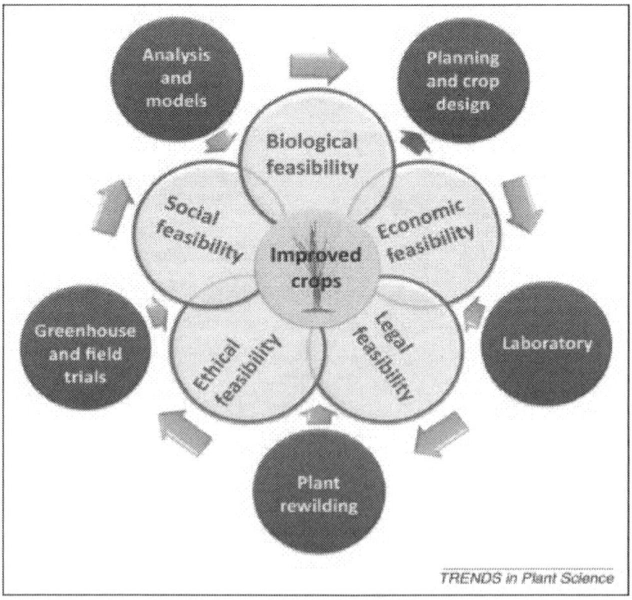

The investigations of Abel et al., 1986 during the 1980s showed that change of tobacco plants (Nicotiana tabacum L.) to communicate the coat protein of tobacco mosaic infection made the plants impervious to this infection. Comparably in potato, potato infection Y (PVY, an individual from the variety Potyvirus; family Potyviridae), potato leaf roll infection (PLRV, an individual from the sort Polerovirus; family Luteoviridae) and potato infection X (PVX, an individual from the class Potexvirus; family Alphaflexiviridae) are the most widely recognized and wrecking potato infections around the world (Kreuze& Valkonen, 2017).

To take care of this issue, infection determined opposition was designed to both PVY and PVX in potato cv. Chestnut Burbank, addressing just the subsequent model (after tobacco) of hereditarily designed infection opposition in crop plants (Lawson et al., 1990). Utilization of infection safe plants is the best and practical method for moderating misfortunes brought about by plant infections. One limit of safe cultivars is the inevitable breakdown of obstruction inferable from the development of another viral strain or species (Kumar et al., 2017). Then again, the utilization of pesticides to control bug vectors is exorbitant and causes hurtful natural outcomes. In this way, taking advantage of methodologies that give sturdy and wide range obstruction is significant in future.

Fungal Resistance

Fungal pathogens cause several important diseases in crop plants. For many years, of fungicides application was the only effective strategy for their management. In present time, considerable progress has been made in identification and cloning of genes involved in plant defense responses.

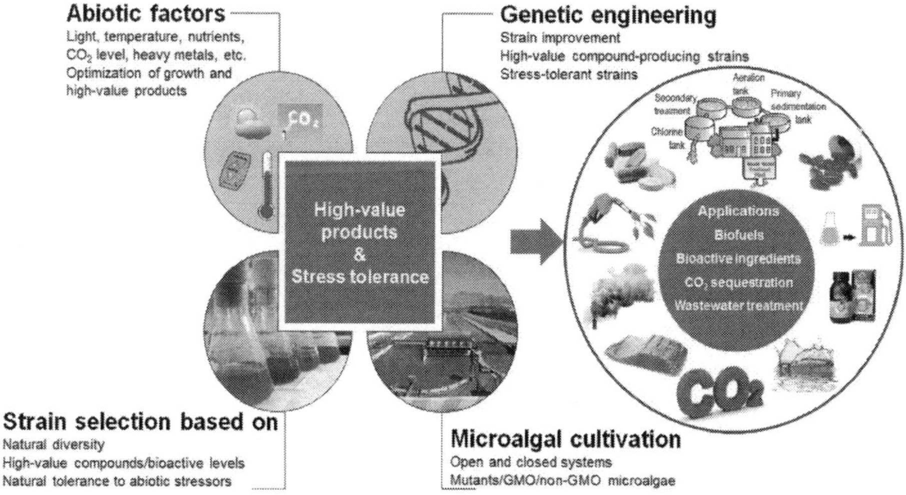

With the aid of plant molecular biology and biotechnology, a large number of antifungal proteins and peptides have been isolated and assessed through in vitro bioassays (Culture and Islam, 2014). An important trend therefore is the technical advance that is made to construct cassettes that contain multiple traits.

Already, this is feasible to a certain extent, as has been demonstrated with gene stacks that contain three NLRs that recognise P. infesting. For many crops the reservoir of cloned resistance genes is still limited. However, another trend is that new affordable sequencing technologies combined with bioinformatics approaches allow ever faster identification of causal resistance genes. On these successes, RNAi has been explored as a strategy to control fungi and oomycetes as well, and initial patent applications for methods to control fungi using RNAi were made as early as 2006. Significant effects have been observed in Fusarium species by targeting the cytochrome P450, family 51 (Cyp51) genes that underlie the azole fungicide target sterol 14α-demethylase with host-induced gene silencing (HIGS) (Koch et al., 2013).

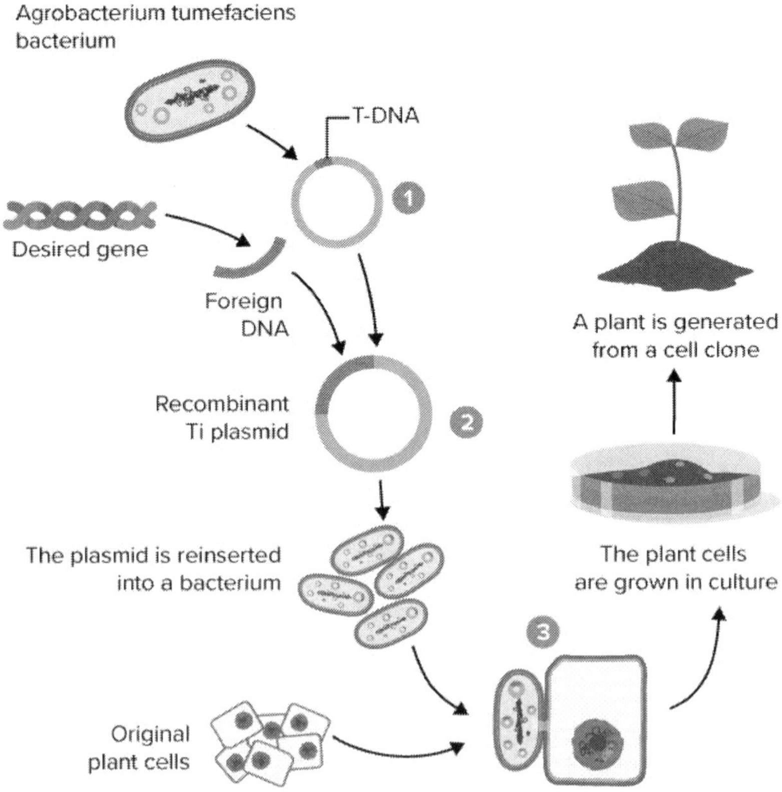

Plant genetic manipulation through expression of either new proteins from foreign organisms or the overexpression of a part of their own defensive arsenal for disease resistance has become a reality (Kumar, 2001).

Bacterial Resistance

The number of different interaction combinations between bacteria and plants and the diversity of types of bacteria with respect to mode of pathogenesis indicate that plants possess various mechanisms for resistance to bacteria. Multiplication of most bacterial pathogens may occur for a period of time in both resistant and susceptible hosts; following this initial interaction, the plant tissue responds in a predictable manner (Klement and Goodman, 1967). The presence and release of phenolics and their oxidation products in diseased plant tissues has stimulated a host of studies attempting to prove that specific phenolics are responsible for resistance to bacteria as well as to other pathogens (Farkas and Kiraaly, 1962). The hydrolysis or oxidation of arbutin may be cited as one of the biochemical processes that provide resistance against the fire-blight pathogen, Erwinia amylovora. Subsequent examination of the hostresponse systems would permit identification of the specific bacteriostatic or bactericidal compounds that are synthesized or released from bound form (Kumar, 2001). Recent evidence that specific protein components from bacterial cells induce generalized resistance reactions indicates good prospects for rapid progress in understanding disease resistance to bacteria.

Herbicide Tolerant

For a long time, researchers and ranchers have realized that herbicide resistance can be moved starting with one plant then onto the next through crossbreeding, in cultivated crops and in wild plants. Individuals have been noticing, examining, and dealing with the exchange of herbicide capacity to bear quite a while some time before the methods of present day biotechnology were utilized to hereditarily change plants to have these qualities.

The GE attribute giving resistance to in-edit utilization of the herbicide glyphosate was presented in soy and canola in 1996 and, in cotton in 1997, altering rural practices for these yields. The presentation of a transgenic glyphosate-open minded (GT) soybean (Gathering Ready® or RR soybean) in 1996 changed farming and empowered a new-use design for glyphosate-based herbicides (Azhakanandam et al., 2015). RR soybean is completely open minded to glyphosate. In this manner, glyphosate can be applied "in crop" as a post-emanant herbicide to control weeds without crop injury.

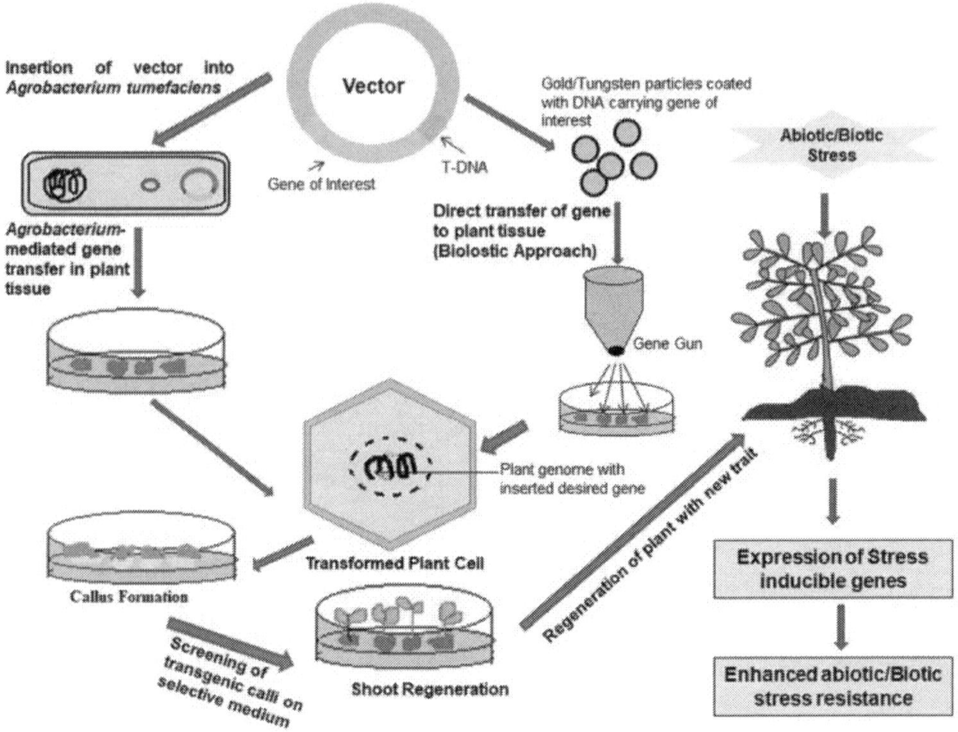

In horticultural regions, where atrazine has been broadly utilized, atrazine safe biotypes of many weed species have showed up. Obstruction was viewed as maternally acquired and was related with changes in the psbA quality (Botterman and Leemans, 1988). Glyphosate-open minded cell societies of Corydalis sempervirens and Petunia hybridawere created after choice on the herbicide (Amrhein et al., 1983). An extensive variety of harvest species has been changed to present protection from herbicides, for example, glyphosate, bromoxynil and glufosinate, and so on. Herbicide-open minded crops are under development in nations, like USA and Canada.

Abiotic Stress

safe During the most recent fifty years, it has been shown that abiotic stresses impact plant development and harvest creation to extraordinary degree, and trim yields have clearly deteriorated or diminished in financially significant harvests, where just high data sources guarantees exceptional returns. Dry spell prompts changes in calcium particle levels, which actuates calcium-subordinate protein kinases (CDPKs) by means of calmodulin-like space. The actuated CDPKs manages downstream parts of calcium flagging. For example, OsCPK4 overexpressing rice plants show expanded water-holding limit under dry spell or salt pressure (Campo et al., 2014). Likewise, overexpression of SNAC1, OsNAC10 and OsNAC5 driven by a root-explicit advertiser RCc3 brought about expanded dry spell opposition under field conditions (Jeong et al., 2010).

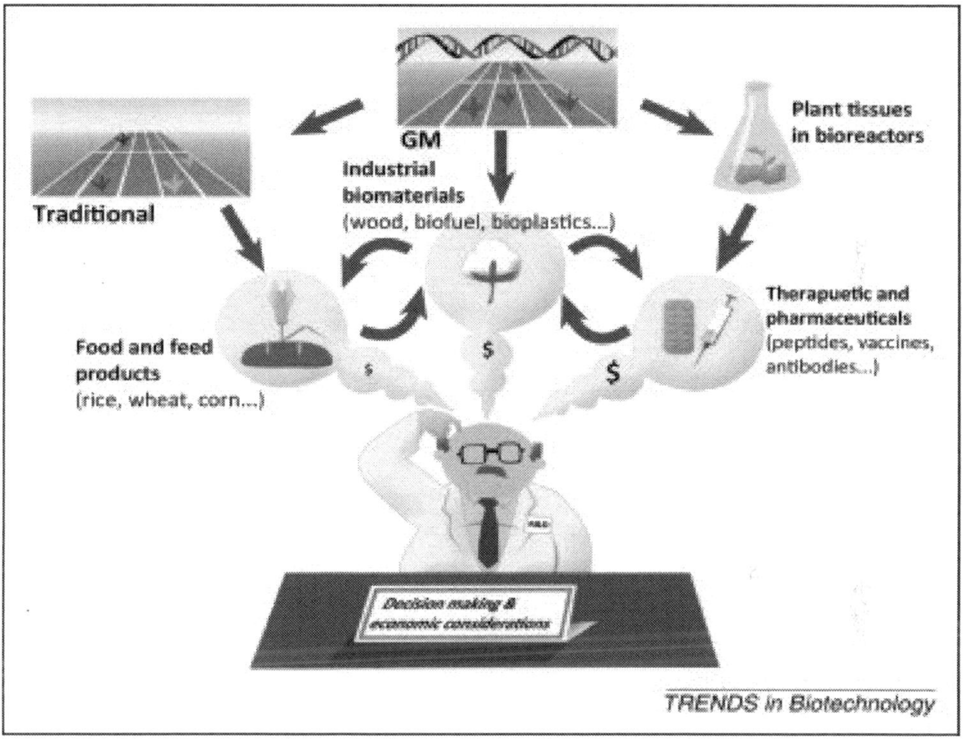

Transcriptional controllers, like HSFs, WRKY, Zat and MBF1c, a transcriptional controller of DREB qualities are enacted to manage articulation of HSPs and other intensity stress reaction qualities (Suzuki et al., 2011). A few digestion related proteins, including starch digestion chemicals, for example, phosphogluconate dehydrogenase, NADPspecific isocitrate dehydrogenase, fructokinase, cytoplasmic malate dehydrogenase, pyruvate orthophosphate dikinase antecedents (PPDK), aconitate hydratase, glycine dehydrogenase and enolase, have likewise been accounted for to be initiated during cold pressure (Lee et al., 2009). Dehydration responsive record factors intercede record of a few qualities in light of cold and water pressure (Kumar, 2001). Today the significance of harvest protection from such ecological perils is probably going to increment further as the scope of climate in which yields are developed, extends and the rate of outrageous weather patterns increments.

Quality Improvement

Dietary nature of the food sources we eat is quite possibly of the main concern, particularly in the creating scene. Plant biotechnology gives massive degree to further develop the food quality concerning proteins, amino acids, nutrients, oil and starch for human wellbeing and prosperity. As far as plant proteins there are two significant ways to deal with work on the wholesome quality: (I) alteration of the amino corrosive creation of the plant proteins. Furthermore (ii) presentation of transgenes that encode proteins of high nourishment esteem. Seed capacity protein (2s) quality (AIIIA I) segregated from Amaranthus is a decent contender for presentation into crop plants as the protein has even amino corrosive structure (Römer et al., 2000). Quality encoding a human milk protein p-casein was communicated in transgenic potato heavily influenced by an auxininducible advertiser (Chong et al., 1997).

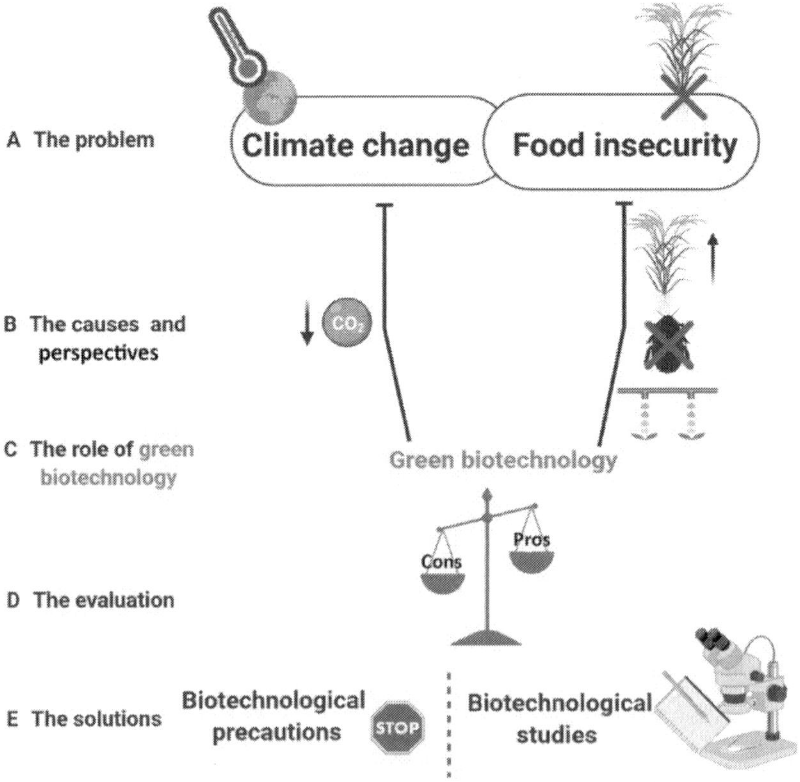

These discoveries open the way for reconstitution of human milk proteins in plant food sources. The nourishing wellbeing and prosperity of people are subject to establish food varieties containing nutrients, minerals and phytochemicals. For example, iron is a significant component engaged with cell processes. the lack of which influences human wellbeing in many pieces of the creating scene (Kumar, 2001). A quality encoding soybean ferritin (ironstorage protein) was presented in rice heavily influenced by glutelin advertiser for the declaration of protein in a seed-explicit way (Goto et al., 1999). The iron substance of the transgenic rice seeds was three-crease higher than that of their untransformed partners. Rice, the significant staple food contains, neitherβ-carotene (provitamin A) nor its C40 carotenoid forerunners. A quality encoding phytoene synthase from daffodil was integrated into rice in an endosperm-explicit way. The transgenic plants gathered phytoene, a key middle in provitamin A biosynthesis in the seed (Burkhardt et al., 1997).

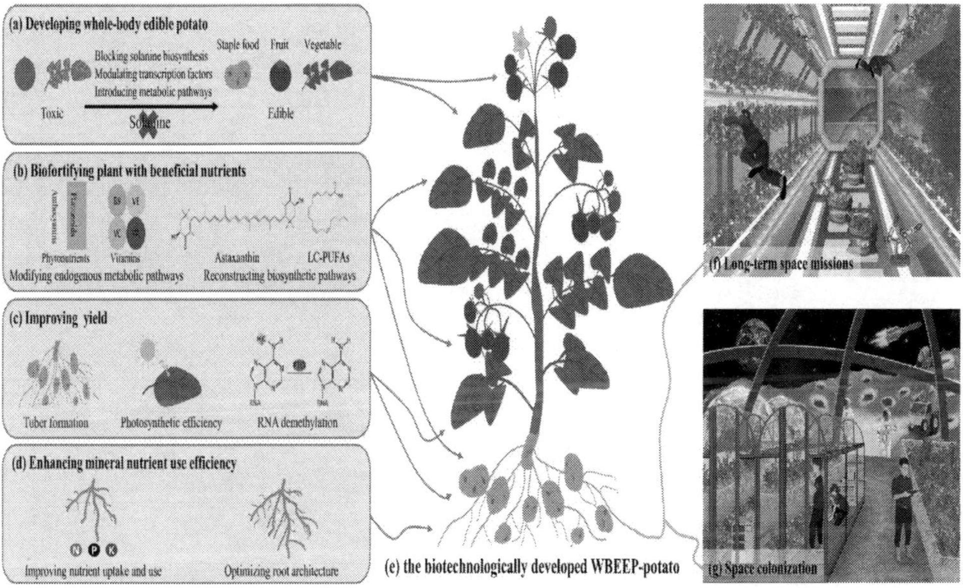

Transgenic oil crops delivering high seed stearic corrosive level give an option in contrast to modern creation of immersed unsaturated fats. Site-coordinated mutagenesis of the quality encoding acyl transporter protein (ACP) thioesterase from Garcinia mangostanaand articulation of this changed compound in canolain a seedspecific way came about in transgenic plants that gather 55-68% more stearatethan the plants communicating the wild-type chemical

Post-harvest

Characteristic Characters that decide the reasonability and capacity life of plant items (organic products, vegetables, blossoms and tubers)

after gather are of high financial significance. For broadening the post-gather life of verdant vegetables, there is a serious need to zero in on the moment perception of these occasions that happen in normal leaves during senescence. It has been accounted for that cytokinins can defer leaf senescence and around then, there is a drop in endogenous ethylene levels (van Staden, 1989). As a matter of fact, tomatoes controlled for postponed maturing are the main hereditarily changed plant items to be showcased in USA (FlavrSavr), in 1994 (Kumar, 2001). Maturing in leafy foods in blossoms can be postponed by antisense articulation of qualities associated with gelatin digestion or ethylene biosynthesis (Kumar, 2001). It has been accounted for that the time span of usability of transgenic organic products was last as long as 60 days at room temperature with practically no adjustment of their hardness and variety (Broil, 2004). In another report antisense, transgenic lines of tomato have likewise been created by adjusting ethylene biosynthesis with an enemy of ACO quality (Schaller, 2007). In future, the experiences pathway investigation of harvest plants through transcriptomics information examination might investigate the genuine digestion of yield plants will be modified to deliver new/further developed assortments or species that are lenient to ecological burdens (Shukla et al., 2015). Control of maturing process in tropical natural products, for example, mango and banana hold extraordinary commitment in not so distant future.

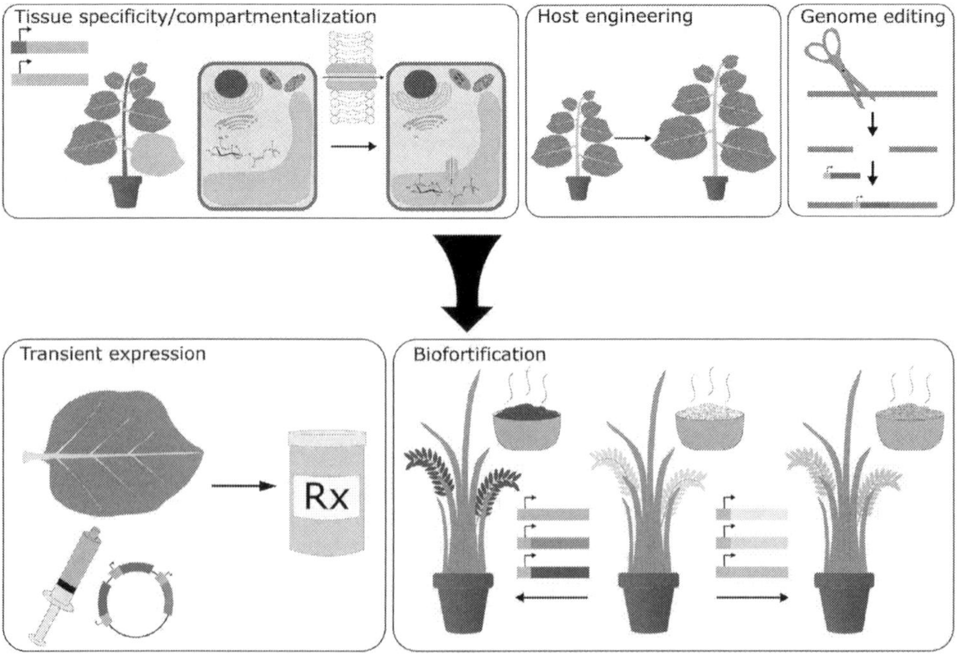

Floriculture

Sub-atomic methodologies are presently being utilized to present beneficial qualities, for example, variety, shape, plant design and jar life to satisfy purchasers' need for curiosity.

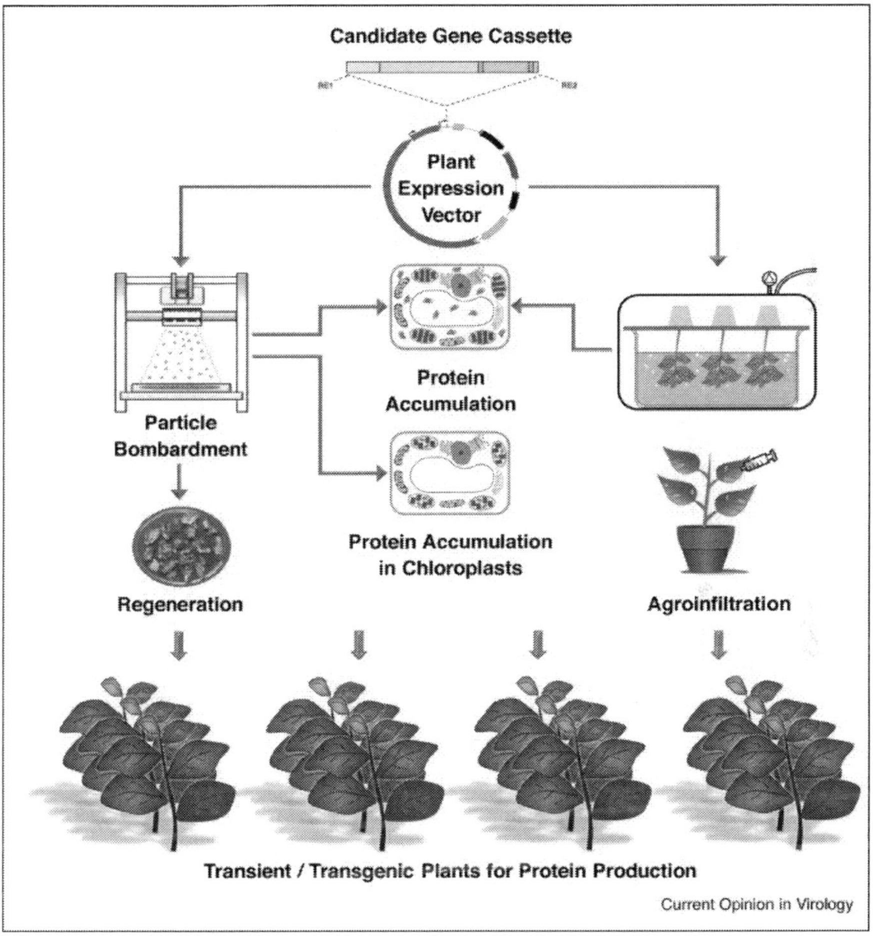

The principal utilization of quality innovation to change blossom tone was accounted for in petunia (Grusak and DellaPenna, 1999). Articulation of the maize dihydroflavonol-4-reductase quality (dfr) in petunia prompts the presence of pale pink to block or salmon red pelargonidin shade and varieties in aggregates. Soliman et al., 2014 concentrated on the impact of gamma radiation on in vitro transformation acceptance in white petals of chrysanthemum (Chrysanthemum morifolium Ramat cv. Youka). The shortening of blooming time by growing early blossoming cultivars or plants ready to deliver blossoms during long days are significant rearing targets in elaborate plant reproducing. Change intervened by Agrobacterium in Sinning sp. upheld that exogenous LFY overexpression advances early blossoming (Chandler, 2010). In Gerbera, change of GSQUA2 quality sped up blossoming. In lisianthus, change of OMADS1 showed huge decrease in blooming time as well as expanded no. of blossoms in contrast with non-changed ones (Noman et al., 2017). Usage of homeotic qualities that controls blossom advancement, might be especially fascinating in fancy bloom crops in which bloom morphology variety can be attractive (Chandler and Sanchez, 2012). The gamma lighted gloxinia plantlets showed morphological changes, for example, cushy leaves, channel molded leaves, short internodes, change of variety blossoms and twofold bloomed (Miri and Roughani, 2018).

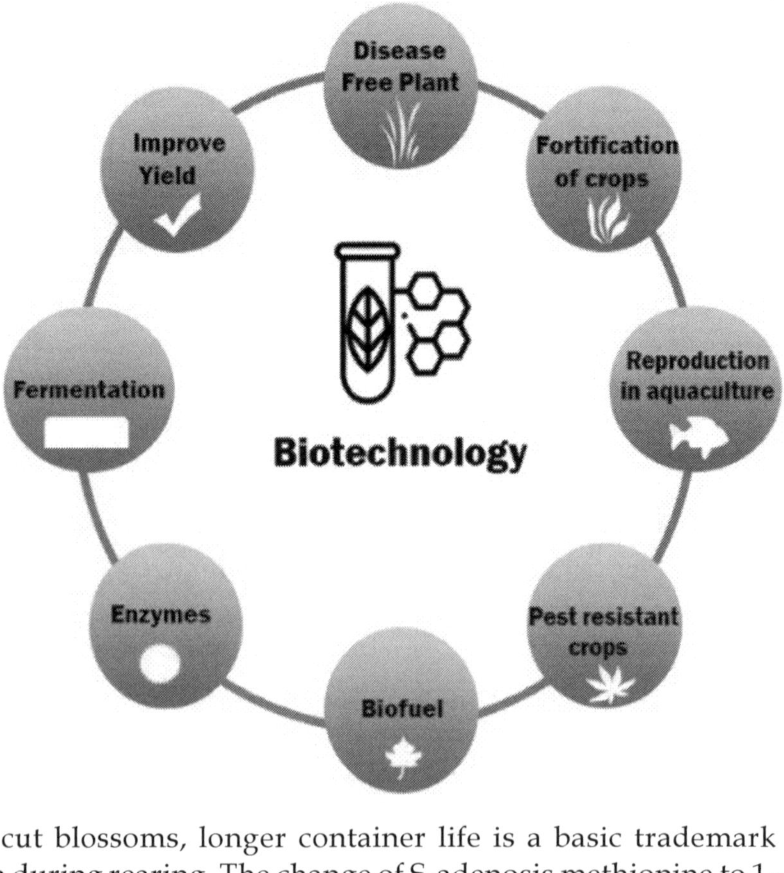

In cut blossoms, longer container life is a basic trademark and is chosen during rearing. The change of S-adenosis methionine to 1 - amino cyclopropane-1 - carboxylic corrosive (ACC) catalysed by ACC synthetase and transformation of ACC to ethylene catalysed by ACC oxidase are the two rate restricting and administrative advances (Kumar, 2001).

Transgenic carnations that contain an enemy of sense ACC oxidase (aco) quality showed low ethylene creation and an obvious postpone in petal senescence (Saving et al., 1995). A few specialists created tetraploids gerbera plants involving in vitro colchicine treatment and found that the ex vitro developing tetraploid plants had longer sprout period with higher container life (Ganpat and Sinniah, 2014). The augmentation of jar life might be because of the way that the tetraploid plants created longer stalks than the diploid.

Phytoremediation

The utilization of plants for recovery of dirtied climate is known as phytoremediation. Over the course of the last 100 years, mining, fabricating and metropolitan exercises have added to broad soil and water defilement. Transgenic Arabidopsis has been created communicating the bacterial quality merB encoding organ mercurial lyase (MerB) with an expect to debase exceptionally poisonous organ mercurial pollutants (e.g., methylmercury) (Cai, 1997).

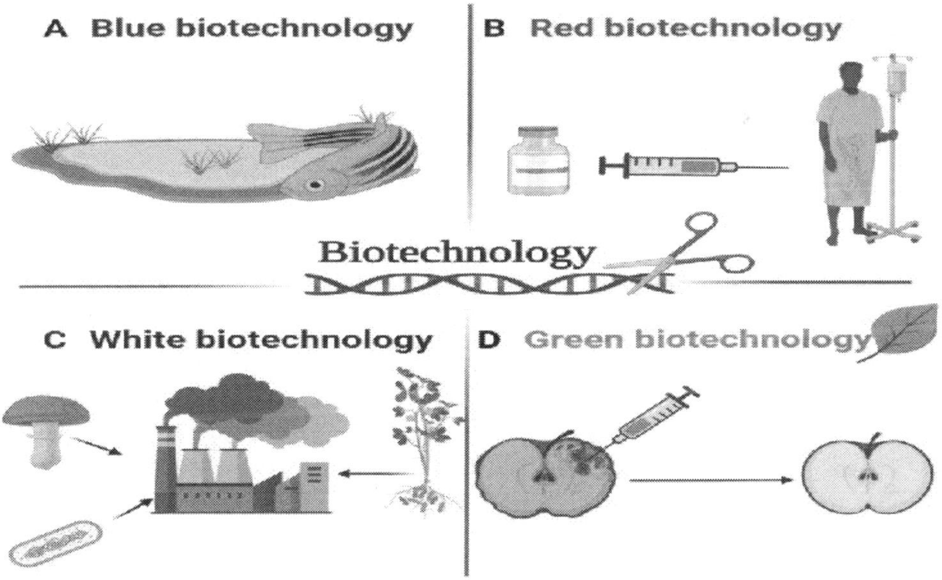

The merrage encodes a compound, mercury reductase that is prepared to do artificially diminishing poisonous ionic mercury to basic mercury which is considerably less harmful and unpredictable. In this way, if these two transgenic plants are developed all the while on the mercury-debased soils, they will actually want to decrease tainting at a significant level (Kumar, 2001).

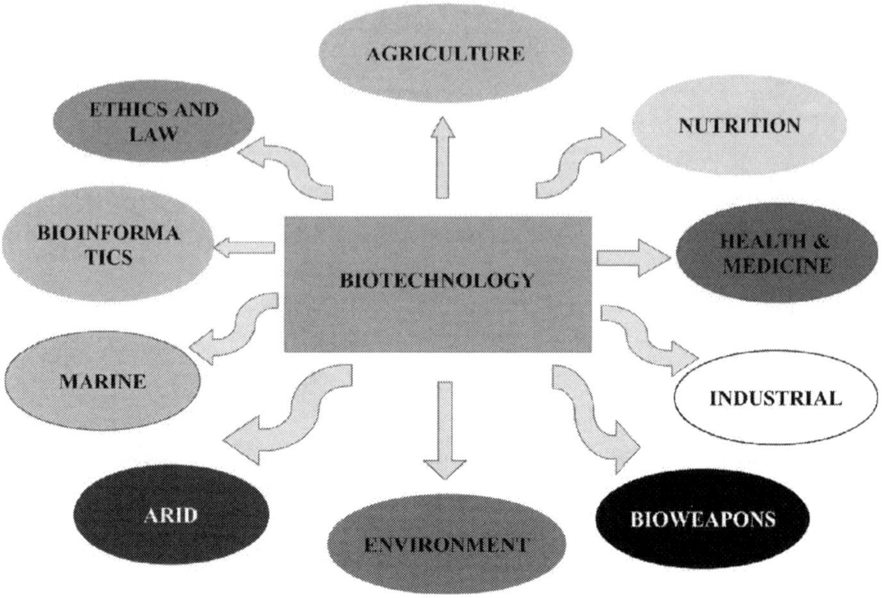

Various transgenic plants have been designed for expanded arsenic (As) resistance and gathering. Over-articulation of qualities that are engaged with the combination of computers or their antecedent GSH fundamentally upgraded arsenic resilience however neglected to essentially improve arsenic gathering (Li et al., 2005). Co-articulation of both γ-ECS and laptops in Arabidopsis delivered a more noteworthy result on arsenic resilience and gathering than over-articulation of either quality alone. Transgenic plants areas of strength for with to arsenic and improved arsenic gathering in the shoots were created by co-communicating two bacterial qualities (Dhankher et al., 2002). At military preparation ranges there is a requirement for remediation of the nitroaromatic explosives, dynamite and RDX (hexahydro1,3,5-trinitro-1,3,5-triazine), to forestall the spread into adjoining networks. Tobacco plants designed with the bacterial quality for a NADPHdependent nitro reductase endure and debase elevated degrees of dynamite (Riot et al., 2006). Furthermore, Arabidopsis plants conveying the xplA quality from Rhodococcus microorganisms are exceptionally impervious to of RDX (Cherian and Oliveira, 2005). RDX can be debased and utilized as a wellspring of nitrogen by a few bacterial strains disengaged from tainted destinations.

Molecular Breeding

Plant Improvement of even the least difficult of qualities needs control of countless qualities, redesign of significant alleles and assurance of their chromosomal area. Atomic procedures are currently accessible to follow the important alleles in isolating populace utilizing hereditarily connected sub-atomic markers. Broad arrangements of hereditarily planned atomic markers, for example, limitation section length polymorphism (RFLP), arbitrarily enhanced polymorphic DNA (RAPD), miniature satellites and intensified piece length polymorphism (AFLP) have been created for some species (Kumar, 2001). Rural yields with added esteem, like those with worked on dietary characteristics, are under nonstop turn of events. Newell-Mcloughlin, 2008 surveyed upgrades to different harvests, remembering improved protein amount and quality for maize, potato, rice, and soybean; expanded nutrient substance (nutrients C, E, or provitamin A) in maize, strawberry, and tomato; expanded carotenoid levels (β-carotene, lycopene, or lutein) in rice, potato, and tomato; expanded flavonoid levels in maize, rice, tomato, and soybean; expanded iron substance in rice and lettuce; and diminished glycoside and salannin levels in potato.

Rice contains minimal normal β-carotene, a forerunner of provitamin A. Brilliant rice was changed to aggregate β-carotene by the addition of exogenous qualities (Ye, 2000). In 2008, two record factors from snapdragons were communicated in tomato to upgrade anthocyanin collection in tomato organic product to levels which are tracked down in blackberries and blueberries (Butelli et al., 2008). In Japan, transgenic strawberries communicating canine interferon-α were marketed and sold as an oral medication from Walk 2014. This is the main illustration of the utilization of a powdered transgenic plant as a medication .The AtIBH1SRDX tobacco plants created multiple times more biomass per unit of development volume, which shows that the upward cultivating that is stacking of numerous racks for plant development, contrasted and wild-type plants (Nagatoshi et al., 2016). In any case, use of quality labeling for significant agronomic attributes is restricted due to trouble in finding firmly connected flanking atomic marker.

Plant Genomics

Plant science research arrives at a milestone with the innovation of sequencing the entire plant genome of specific plant species. Genome sequencing will open skylines to concentrate on plant science, uncommonly portrayal of cell, physiological and formative pathways (Kumar, 2001). An underlying utilization of plant genomics has been to notice quality articulation at a bigger scope.

The methods that have made significant commitments to produce a profile of articulation levels is RNA profiling in view of hybridization of records to varieties of DNA particles bound to a strong help. It is prevalently depicted as DNA chip innovation, and has overcome any barrier between arrangement data and practical genomics (Baldwin et al., 1999). The DNA chip innovation worked with equal securing of huge information for thousands/a great many explicit DNA groupings at a quicker rate through mechanization, trailed by the examination of this information utilizing PC gadgets. As a general rule, the benefit of exhibits is that they give hundreds or thousands of explicit qualities at the same time. In this manner, two general kinds of DNA chips or miniature clusters have been created: DNA part based miniature exhibits and oligonucleotide based miniature exhibits). D'Amico-exhibits can likewise be applied to evaluate populaces of plants for polymorphic 'articulation fingerprints'. These articulation fingerprints then, at that point, can be correlated with a perplexing cycle, for example, dry spell tolerance, to assess new qualities or gatherings of qualities in that plant interaction (Kehoe et al., 1999).

Problems and Future Perspective

Plant biotechnology has made huge examinations in the previous 10 years. Many plant species designed for articulation of various characters are now under broad development in numerous districts of the world. Protection from herbicides, bug bothers and infections has outweighed everything else while presenting the original transgenic crops is trailed by sluggish maturing leafy foods flower characters. The second-generation transgenic plants will have more business ramifications. The frameworks of sub-atomic rearing proposition incredible potential for creation of specific yields which have extra worth.

In any case, crop rearing explicitly for shut development frameworks had been restricted to date regardless of its expected accessibility. Novel cultivars are expected to identify with the difficulties of farming innovation. The significant test in evaluating the exhibition of presented quality in the field of transgenic science is quality quieting, which is a sad finish of a logical undertaking and has prompt commonsense ramifications.

Transgene quieting could be a direct result of presence of different duplicates and hyper methylation of presented quality, and due to presence of homologous successions in different designs in the plant genome. Quality hushing can maybe be to a great extent stayed away from by choice of transgenics containing just single duplicate qualities, as quieting has been many times found to happen when numerous duplicates of the transgene are embedded in the host genome. The tremendous biodiversity and rich germplasm of harvests of Nepal, present before us has incredible difficulties as well as unexpected open doors. Segregation of qualities and advertisers from various living beings is an action which needs more consideration. The capability of utilitarian genomics should be tapped without losing a lot of time. Patent regulations and systems to safeguard native endeavors and discoveries should be created and smoothed out.

Conclusion

The unique time generally offers a chance to ponder human movement in a specific discipline and to figure out a future methodology. Specialists continually test the previous events to learn examples that could end up being useful to in the securing of new information or for the further advancement of fitting innovation resulting from it. Science and innovation can't be confined on the planet, so specialists are supposed to act as per the changing worldwide situation wherein they live. Uses of biotechnology in the hereditary improvement of harvests have been an extraordinary errand in plant biotechnology to be directed in coming many years.

Barely any many years sooner, plant biotechnology was depended upon not many applications just, for example, tissue culture, recombinant DNA innovation and monoclonal antibodies. Today, change, and marker-helped choice and rearing are only a couple of the instances of the uses of biotechnology which are found according to require by the harvest improvement program. Plant biotechnology is encircled by a large number of logical instruments and procedures for screening and hereditary control of plants to foster gainful or valuable plant/ plant items. The capability of these devices and procedures could be additionally helped significantly by nanotechnological intercessions. Utilization of nanomaterials as vehicles for quality or proteins and nanocrystals for high-goal imaging of the plant cell and organs could be one more field to investigate by the scientists for making better curiosity in food crops. We can reason that plant biotechnology is an integral asset for the improvement of new plant qualities and assortments and such new assortments should be created for an enormous scope to make business progress and to fulfill the interest from cultivators and buyers in general.

Made in the USA
Monee, IL
12 November 2026

d3c76d20-527b-40c2-a490-f25746989f2eR01